D1810660

Science, Man and Morals

Science, Man and Morals

BASED UPON THE FREMANTLE LECTURES
DELIVERED IN BALLIOL COLLEGE OXFORD
TRINITY TERM 1963

W. H. THORPE Sc.D., F.R.S.

Fellow of Jesus College Cambridge

GREENWOOD PRESS, PUBLISHERS
WESTPORT, CONNECTICUT

Library of Congress Cataloging in Publication Data

Thorpe, William Homan, 1902-
   Science, man, and morals.

   Reprint of the ed. published by Scientific Book Club,
London.
   Includes bibliographical references and index.
   1. Religion and science--1946-   2. Science and
ethics. I. Title.
[BL240.2.T47 1976]   215   76-14962
ISBN 0-8371-8143-7

First published in 1965 by the Scientific Book Club, London

Reprinted with the permission of W. H. Thorpe

Reprinted in 1976 by Greenwood Press, Inc.

Library of Congress Catalog Card Number 76-14962

ISBN 0-8371-8143-7

Printed in the United States of America

Contents

v

Contents

Contents

'*Il n'y a de création que dans l'imprévisible devenant nécessité.*'
Pierre Boulez *Revue Musicale* 1952

Preface

This book has arisen from a course of lectures delivered in 1963 under a lectureship established at Balliol College by the will of Sir Francis Fremantle. The lecturers are required to discuss the relation of science to Christianity and the Christian life in essence or in its various applications, with particular reference to education or to the sympathetic explanation of differences between disciplines and different religious communities. I felt it a great honour, when I received the invitation of the Master and Fellows of Balliol to be the second lecturer on the Fremantle Foundation. The hospitality of the Society was memorable; and discussions with some of the Fellows and other members of the audience proved very stimulating and contributed materially to improvements and developments in the process of making a book from five lectures. The general organization of the lectures has been followed in the book, but the latter is in many ways very different from the former and, as will be evident, I refer to many sources and developments not yet available when the lectures themselves were given.

It has been the accepted custom, at least since the first popular essays of Sir James Jeans, for Physicists and Astronomers to speak their mind readily on the implications of their studies for philosophy and theology; and the spate of such works has continued unabated. Biologists on the whole have heretofore been more reticent; although there were of course some famous examples who concerned themselves with evolutionary topics about a century ago. This reticence has been due in part no doubt, to the slower momentum of biological studies and the smaller, more specialized scope of much biological work. In recent decades, however, the situation has changed and it is now the life sciences quite as much as the physical sciences that are providing challenge after challenge to our generally accepted ideas and modes of thought about man, his nature and destiny.

In the biological sciences, as of course in the physical sciences also, this is a time of unprecedented ferment. The rate of change has been and is increasing in a bewildering fashion – though of course it follows that many recent conclusions are in fact far more provisional and tentative than the readers of the popular science magazines might imagine. Without any doubt some of the apparent philosophical implications thrown up by the biological sciences today will appear dated or will show a very different complexion to our successors in the year A.D. 2000 (assuming that we have any successors!). But the tentative nature of theories is part of the very stuff of science and is in no sense reason for failing to consider and discuss them from every angle; indeed scientists, philosophres and theologians, fail in their obligations to contemporary thought if they do not do so. Moreover, however problematic and doubtful some of the more esoteric implications of scientific advance may seem, there is no doubt whatever about the practical applications of science. All our horizons are darkened by their menace and at the same time lit by their promise. This is every bit as true of the applications of the biological sciences as of the physical sciences. The ethical problems raised by the population explosion and artificial insemination, by genetics and neurophysiology, and by the social and mental sciences are at least as great as those arising from atomic energy and the H-bomb, from space travel and ultra-sonic flight, from telecommunications, computers and automation. There is no doubt in my mind that several of these developments are as epoch-making for mankind as any that have preceded them. They rank at least as high, if not higher, in importance than the discovery of fire, of agriculture, the development of printing, and the discovery of the wheel.

Preface

The lectures were open lectures and were not addressed primarily to scientists. They were as far as I can judge generally intelligible to the audience and it is my hope that the book may appeal to an equally wide spectrum of readers. Yet I fully realize that some parts of the work are bound to be difficult for non-specialists: in these days of fragmentation and specialization in the ever growing range of scientific investigation, communication between scientists is becoming ever more difficult and such difficulties are greatly increased when one is attempting to speak on these subjects to those outside the scientific field. There is, I am sure, one part of the book which is likely to be found too severely technical for the general reader, namely the first section of Chapter 2. I suggest that those who find difficulties here omit this part, at least at first reading. For their benefit I can sum up this section, perhaps not too inadequately, by saying that I believe it provides new and important scientific evidence for the famous statement by Gowland Hopkins, one of the founding-fathers of biochemistry and discovery of vitamins, who described the origin of life as 'The most improbable and most significant event in the history of the Universe'. In many respects this is a central theme of the whole book. One can almost define the subject as 'The Process of Creation'. The other main theme, if indeed it is a distinct theme at all and not simply an aspect of the first, is the impressive and growing understanding of the unitary foundation underlying all experience – scientific, artistic and religious. This I regard as a kind of inexorable ground-swell of modern man's awareness of himself and the world. It finds expression in many of the most profound writers of recent times; including philosophers and theologians as far apart as Whitehead, Berdayeff, Collingwood and some of the Existentialist school through Teilhard de Chardin to scientists such as Eddington, Schrödinger, Hinshelwood and Polanyi. The theme is indeed tremendous. Perhaps my temerity in approaching it will, if it does nothing else, stimulate others better qualified and equipped than I to do likewise.

xi

Preface

It will at once be evident to the reader that in producing this book I have been particularly indebted to and stimulated by three contemporary writers, all of them scientists of great distinction. They are T. Dobzhansky, C. H. Waddington and Michael Polanyi. To Dobzhansky's great book *Mankind Evolving* I have been particularly indebted. My own thought is in close sympathy with his and his work has been a constant source of inspiration from which I have frequently quoted. The writings of Professor Waddington I have also found stimulating at every turn, though my response is perhaps more often critical than in agreement. To Professor Michael Polanyi, now a Fellow of Merton College, Oxford, I am greatly indebted; not merely for his book *Personal Knowledge* and for many published papers and lectures, but above all for some prolonged discussions, profound as they were delightful, and for giving me access to much work as yet unpublished. But great as is my indebtedness to him I would not like him to feel in any way saddled with the conclusions here expressed. Though I do not doubt that he would agree with much, some I am sure he would disagree with or at least wish to express very differently.

In the course of rewriting and revising the book I have been indebted to a number of friends and colleagues, notably Dr Malcolm Dixon F.R.S., Fellow of King's College, Cambridge; Professor Donald MacKay, Professor of Communication in the University of Keele; Dr Clifford Evans, Fellow of St John's College, Cambridge; Dr Martin Wells, Fellow of Churchill College, Cambridge and Miss Barbara Lade of my own laboratory.

Jesus College,   W. H. THORPE
Cambridge
December 1964

CHAPTER I

The Nature of Life and the Idea of Creation

LIVING AND NON-LIVING

Those at school forty to fifty years ago were not troubled by any doubts about living things being indistinguishable from non-living. We were told that crystals resembled living things in that, under conditions which could easily be contrived in the elementary laboratory, some of them would grow in a predictable way which suggests the growth of living organisms. But no one bothered to take this very seriously – for it seemed self-evident that a crystal of alum or copper sulphate growing in its mother liquor had only the most superficial resemblance to a living thing. The virus, though first discovered before the turn of the century,[1] was not yet considered a proper subject for the student of elementary science.

Now all is different. As biological knowledge has accumulated, so has the meaning of the ordinary English words 'life' and 'living' become more and more indefinite. Some at least of the features which are usually regarded as characteristic of life are now detectable in many entities which, at first sight, we should certainly not have thought of as living organisms. Such are not only the viruses, bacteriophage and certain enzymes, but also giant molecules, including among these certain proteins.

Now there are two characteristics of even the simplest living things which do seem to offer a distinction between biology and the physical sciences.[2] The first is that life seems to be a highly

dramatic affair which involves the maintenance of a kind of steady
state as a result of continual active interchange with the environment. The organism is continually extracting what it needs from
its surroundings and returning to these surroundings waste pro-
ducts of one kind and another – that is to say, it is an open system,
maintaining an equilibrium or steady state as compared to the
closed, inert, non-metabolic systems of the chemist and classical
physicist. Secondly, another great feature is that living organisms
must be thought of simultaneously on a number of different time-
scales. First, there are the continual moment-to-moment activities
of a living thing, as a going concern. In addition to these there are
the slower processes by which an egg develops into an adult and,
on a yet longer time-scale, there are the processes of heredity by
which the characteristics of organisms are passed on from one
generation to the next. Finally, on a time-scale of hundreds of
generations, there is the slow process of evolution, as a result of
which the character of the individuals in a population gradually
changes, and populations may be split up into two or more
different species. As I shall argue, all these processes appear
strikingly non-random: in fact, 'directive'. The biologist has to
organize his studies so as to cover appropriately all these different
activities with their different time-scales; and this at once gives to
his discipline a complexity unknown in the physical sciences. The
second of these points will be discussed below, but it is convenient
to consider the first more carefully now.

Right at the start, modern biology confronts us with a dilemma:
life appears to be an activity, a kind of dynamic equilibrium, but
looked at closely a difficulty arises. In recent years, techniques have
been developed which enable us to induce, almost instantaneously,
intense cold. In this way micro-organisms (and even sometimes
creatures as high in the scale as the smaller mammals) can be deep-
frozen as in a flash, and thawed out again at will at a similar speed.
When this is done, it is found that life is not after all, dependent
upon the continuity of metabolic processes; which certainly stop
completely at these very low temperatures; for organisms thus

treated regain their normal appearance and metabolism on thawing. They 'come to life' again; and may continue to function in an apparently normal manner as if nothing had happened. This new development recalls the type of research which greatly occupied the attention of learned men in the late eighteenth and early nineteenth centuries, research into what was then known as 'suspended animation' (now anabiosis) due in that case to complete desiccation. The resuscitation of desiccated micro-organisms when they were again provided with moisture astounded the naturalists of those times.[3] What we observe now when animals are deep-frozen with the proper technique recalls very dramatically these old drying-out experiments. As a result, then, of this new development, there seems no doubt that at very low temperatures all the biochemical and physiological reactions – in other words, all metabolism – can be reversibly suspended for considerable periods of time. This at once seems to point to the conclusion that the condition of life is, after all, essentially structural; and the flux of activity which we see, and have so long regarded as of the very nature of the living thing, is, in a way, only secondary. It is as if the deep-frozen organism has the potential vitality which can inexorably result in actual life the moment the right physical conditions are provided for it. This conclusion, that the condition of life is essentially structural, is of profound theoretical importance. Thus there seems no theoretical underlying reason why a new Rip van Winkle might not be deep-frozen for a thousand or indeed a million years, and reawake exactly as he was, unaware that any time had elapsed. Keilin [4] says, 'The concept of life as applied to an organism in the state of anabiosis becomes synonymous with that of the structure which supports all the components of its catalytic systems. Only when the structure is damaged or destroyed does the organism pass from the state of anabiosis or latent life to that of death'. If we accept this it must not be forgotten that 'living' organisms may be temporarily converted into 'organisms' as when bacteria, semen or slime moulds are dehydrated and stored in this condition.

A further point follows from this: if the essential basis upon which life of the organism appears to depend is not after all a process but a micro-structure, we immediately feel more confident (though perhaps unjustifiably) that if man is able to create the right chemical structure and maintain it in the right chemical environment, he will thereby create life and with it whatever mental qualities are appropriate to its degree of organization. I personally shall indeed be surprised if it does not prove possible, within the not very distant future, to produce artificially in the laboratory chemical entities which, most of us would agree, deserve the description 'living'.

But yet, if structure is so fundamental, it is because it enables the characteristic metabolic activity of the living organism as an open system, [5] to proceed.

It is perhaps symptomatic of a big advance in undertaking that recent students of the nature of the life process are seeking to relate the structural and the functional in some more primary, more all-embracing definition of life. Thus the eminent biochemist Szent-Gyorgyi has devoted attention to the origin of chemical energy which enables the living organism to move, whether by muscles, or other kinds of organs serving for movement.[6] He argues that the mobile form of energy which it is necessary to assume in order to explain the movement of muscle is equivalent to electronic excitation energy. The release of this form of energy in a conventional physical system (10^{-9} seconds) is too short; but by changing the direction of the electron spin this time becomes lengthened a million-fold. Thus uncoupling the electron spin is the foundation of the energy on which the whole living world is built. Systems in which this uncoupling can be done, and the appropriate electro-magnetic field created (i.e. in one or other of a very few types of large molecule) provide one of the matrices of life, water being the other. As I understand it, what this amounts to is that the whole activity of life is based upon a process similar to the photosynthesis carried out by green plants, and its reversal. This cannot be described in the classical symbols of chemistry and Szent-Gyorgyi argues that the most essential happenings which provide the energy for living things are not chemical but lie in the field of quantum mechanics. It has been said that he has thus introduced a new subject, namely 'quantum mechanical biology'. But its possible bearings on the problems of 'life' and emergence are still unclear. (See however p. 27 below.)

Now although these developments show that analyses to ever finer levels of structures are revealing new facts of quite fundamental importance concerning the basic conditions of life; yet it is becoming clearer that the physics and chemistry of today, which seem to be providing new insights into the nature of life, are vastly different from the physics and chemistry of a generation ago – sciences to which the biologists of that time turned so hopefully, namely, the physics of matter *en masse* and the chemistry of small molecules. Extension is in both directions, i.e. to smaller and to larger. As Picken [7] says, 'we can see already that the physics and chemistry are going to be those of ordered aggregates far larger than molecules. The extent to which the concepts of chemistry and of physical chemistry have been modified sometimes almost out of existence – as in the concepts of molecule – as a result of contact with biological material is of the greatest historical importance.' In some respects, as Picken forecast, we seem now often to be starting with concepts derived from biological systems and working back into the physical sciences. Thus when Schrö-dinger [8] calls the gene or the chromosome an a-periodic crystal, or when the insect organism [9] is considered as a gigantic molecule, this is not (as Picken shows) the abuse of language that it may appear at first sight, but may in fact be enshrining important insights. Can we not speak of the isolated virus and the deep-frozen protozoon as 'organisms' in the basic sense; reserving the term 'living organisms' for those with the metabolic turnover?

CAUSATION AND THE CREATION OF LIFE

In recent developments of biological thought, we seem then to see a new development in the way of explanation based upon atomism and yet at the same time a tendency for the organismal approach to take over some at least of the fields of physics and chemistry which a few generations ago would have been regarded solely as the sphere of the atomic. Are we in fact witnessing the birth of a new kind of mechanistic explanation and at the same time a new kind of vitalism each more appropriate and more embracing than the old kinds, and if so, how does this affect our ideas of creation and the argument from design?

In the year in which *On The Origin of Species* was published, William H. Gillespie wrote a little book [10] in which he tried to explain how the geological picture of whole genealogies of extinct organisms such as the giant reptiles, which succeeded for millions of years in the evolutionary struggle by specializing as ruthless predators upon their weaker neighbours, could be understood in terms of Christian theology. He starts with the then orthodox view that the lower animals suffer and die because of man's sin. On the basis of this he appears to doubt the very evidences of geology because he could not believe that a benevolent Creator would have created such frightful monsters. But although he makes the attempt to deny geology the status which qualifies it to establish any facts other than those with purely physical implications, in the end his courage fails him and in a supplementary paper (pp. 78–83) he argues that perhaps after all there were such monsters, but they were not created directly by God. They were originally innocent creatures led astray by the devil; or, perhaps, like the Gadarene swine, they were actually animal bodies inhabited by the spirits of demons. This might, he thought (p. 83), explain why the Bible contains the story of the Gadarene swine, which has been a stumbling block to so many. P. H. Gosse in his famous, one might say notorious, book *Omphalos* [11] admitted freely all the evidence of geology in favour of the antiquity of the world but got out of the difficulty in a manner all his own. He argued that when the creation took place everything was constructed *as if* it had a past history. Thus Adam and Eve had navels just as if they had been born in the ordinary way; and similarly everything

6

else was created as if it had grown. On this view the rocks were filled with fossils just as they would have been if they had been due to volcanic action or to sedimentary deposits and, as Bertrand Russell says [12] we might all have come into existence five minutes ago, provided with ready-made memories, holes in our socks and hair that needed cutting! But however strongly entrenched the argument from design, nobody could believe that the universe was designed like that; and poor Gosse, to his immense dismay and chagrin, was laughed at by both scientists and theologians. But is there a modern form of the argument from design that is reasonably plausible and respectable? Students of the history of science, notably Basil Willey, have pointed out that both St Thomas Aquinas and St Augustine held that 'in those first days God made creatures primarily and *causaliter*'. This meant that he created not the finished creatures but potencies, such that the earth would gradually give rise to them. Aquinas continues that after the primary creation God 'then rested from His work and yet after that by His superintendence of things created He works even to this day in the work of propagation'. I agree, of course, with Waddington [13] when he says 'acceptance of evolution does make it impossible to believe in a God who operates in the way in which a literal interpretation of Genesis suggests; but ... interpretations do not have to be literal'. Thus it is possible to have a form of the argument from design in which the design consists solely in the establishment of the initial conditions of the universe and all else follows from this single first act of creation. But if we take this view then clearly the whole of the cosmos, and the evolution story, was determined in some primordial nebula and I shall try to show in the next chapter that this is untenable, *as a purely physical hypothesis*. Sir Ronald Fisher, once argued [14] that simple causation of a strictly deterministic kind cannot produce results which can legitimately be described as creative – that is to say any creativeness manifest must all have been present in the initial act. His view is that the evolutionary process is and can be creative simply because at bottom the universe is not physically deterministic; [15] and in a strictly deterministic universe, unin-fluenced by chance, there would be no room for anything which could be regarded as essentially new and thus as a result of a creative process. He argues that natural causation has a creative aspect because it has acasual aspect. These are the back and front of the same quality. And it is only because the casual processes in the actual world have an essentially chance element in them that we can find any situations to which the concepts of creation can properly be applied. Looking back at the cause we can recognize it as creative. It has brought about something which could not have been predicted, something which cannot be referred back to antecedent events. Looking forward to the same 'cause' as a future event, there is in it something which we can recognize as casual. It is viewed thus like the result of a game of chance; we can imagine ourselves able to foresee its forms and to state in advance the probability that each will occur. We can no longer imagine ourselves capable of fore-seeing just which of them will occur. But this view is based ulti-mately on a conclusion of physics about which there is still much controversy.

What about the vitalistic approach? The old vitalism argued that living systems – as in the experiments of Driesch, in which when an egg is cut into fragments, each fragment develops into a complete adult – could not be explained without invoking the activity of some non-material and non-mechanical 'whole making' active agent. He called this agent an 'entelechy'. But there is another form of vitalism according to which all phenomena are potentially explicable in terms of physics and chemistry except the phenomenon of self-awareness which in essence we can only observe subjectively and that this is of an essentially different nature which cannot be subsumed under any view of the universe as primarily a physical system. The old vitalism has gone un-lamented. The new vitalism, as Waddington says, [16] is difficult to refute; but he points to a fact which, in his view, leads to the evaporation of the whole controversy. Here again he is I believe making the common mistake of (under the plea of ignorance as to what material components really are) putting what is to be explained into the concepts which are to serve for explanation. This is the error discussed below under Polanyi's term 'pseudo-substitution'. Nevertheless he makes an [17] historical point of interest. 'Objective vitalism amounted to the assertion that living things do not behave as though they were nothing but mechanisms constructed of mere material components; but this presupposes that one knows what mere material components are and what kind of mechanisms they can be built into. In late Victorian times, in the heyday of the physics of Newton and Faraday, and the chemistry of Dalton, people had a surprising confidence that they really knew what the world of matter consists of . . . it was one of the only too numberous aberrations of Victorian self-confidence to think that they knew enough about atoms to provide vitalists and mechanists with anything to argue about.'

It was perhaps the greatest of the great contributions of Whitehead to show in his *Philosophy of Organism* that even the antithesis atomism *versus* continuism was a false one, and that perhaps after all it is organicism that is fundamental. If this is so, then fundamental biology is the study of genes, cells, tissues and whole organisms and their behaviour. The most highly developed relationships that living things exhibit are just as fundamental, perhaps more fundamental for science, than the idea of the ultimate atomic unit. He summed all this up in his famous aphorism, 'Biology is the study of the larger organisms and physics the study of the smaller ones.' But here too we must be on our guard against lightly supposing that the 'organicism' of the smaller units is in any way such as to account for that of the larger.

It seems, however, that the argument from design is tenable; indeed, in the sense that the physical environment is fit for 'life', inescapable. This has been persuasively argued by Pantin[18]: 'The organism is thus built up of standard parts with unique properties. The older conceptions of evolutionary morphology stressed the graded adaptation of which the organism is capable, just as putty can be moulded to any desired shape. But the matters we have discussed lead us rather to consider the organism as more like a model made from a child's engineering constructional set: a set consisting of standard parts with unique properties, of strips, plates and wheels which can be utilized for various functional objectives, such as cranes and locomotives. Models made from such a set can in certain respects show graded adaptability, when the form of the model depends on a statistically large number of parts. But they also show certain severe limitations dependent upon the restricted properties of the standard parts of the set.'

He goes on to argue that any functional problem must be met by one or other of a few possible kinds of solution. If we want a bridge, it must be a suspension bridge, or a cantilever bridge, and so on. In the problems of design there are in fact three elements: firstly, the classes possible in this universe; secondly the unique properties of the materials available; and thirdly, the engineer or architect who, by selecting the class of solution, and by utilizing the properties of his materials achieves the job. Pantin continues: 'We can apply these ideas to the construction of the living organism. Like all material structures they must conform to certain constructional principles. The standard parts available for the construction of organisms are the units of matter and energy which can exist only in certain possible configurations. Like the engineer, natural selection takes third place by giving reality to one or other of a series of possible structural solutions with the materials available. But the fact remains that we have arrived back at the eighteenth-century conception of an ideal plan as an essential constituent of organic design. . . . In the child's constructional model of a crane we discern not one principle of design but at least two. For there is the design of the set of parts so that they shall build such things as cranes as well as the special design of the crane. In the living organism we can ascribe the apparent design of its immediately adaptive features to natural selection. Can we discern design in the properties of the units which make such an organism possible? These properties of the units are not the result of selection in the Darwinian sense. And if we see design in them we must say with du Bois-Reymond:

'... whoever gives only his little finger to teleology
Will inevitably arrive at Paley's discarded "Natural Theology".
Natural selection bowed Paley and his argument from design out of the front door in 1859 and here he would come climbing in through the back window saying that he owns the title deeds of the whole estate!'

Pantin's argument seems perhaps to imply a 'first cause' as evinced in the basic physical design of the universe and to that extent to be consonant with Deism. It appears secondly to rely on natural selection as the creative agency and therefore not to re-quire a 'creator'; thus as far as it proves successful it does not lead to any form of Theism. However, as we shall see, a system con-sisting of a cosmic basis, describable in purely physical terms, plus natural selection, runs into enormous difficulties when used as the complete framework for an 'explanation' of the universe. But what in any case can we mean by creation? The term 'creation' can be held to imply merely that new and unpredictable things arise in the course of ages. In this sense natural selection is creative and the idea of any other kind of creative agency (a creator) is only doubtfully involved. But if we detect an overall trend towards perfection or towards a 'higher' state – if, that is, we have even the dimmest evidence for a plan – then the hypothesis of a creator begins to appear plausible. And if we think we see a basic design or series of designs, then I believe the creator concept becomes hard to avoid. And I shall try to show that basic design of materials and conditions, impressive though it is, is again insufficient to account for the modern scientific world picture and it is plausible to argue that 'design' must have been operative at innumerable levels in the evolutionary story.

It appears to me, nevertheless, that the facts of submolecular biology and quantum mechanical biology to which I alluded above have also made Pantin's form of argument from design still more compelling and extend its range. For Pantin, the success of natural selection in producing so much that appears to us as designed is dependent upon the fact that 'the universe is such that the right materials are available and appropriate engineering solutions are possible'. But this fitness of the conditions of existence would have to account for such remarkable features as the emergence of individual consciousness; features which seem to elude explanation on this basis alone – even granting that the unique properties of carbon, hydrogen and oxygen,[19] are fully accounted for in physical terms. Pantin, like so many other present day biologists, will not follow Eddington in considering that the existence of a design implies a designer because he feels there is no way of proving whether this is true or not. But is not the onus first on the scientist (if he is one who is attempting a complete synthesis) to show that a creator is an unnecessary hypothesis? And is it conceivable that he can do this?

Recently Dobzhansky [20] has said, 'by far the most impressive demonstration of the unity of life is given by the discovery that the genetic code throughout the living world is composed of only four letters of the genetic alphabet. All the biological evolution extending over a period of some two billion years has occurred on the level of genetic "words" and "sentences"; no new "letters" have been added or, as far as is known, lost. The simplest interpretation of this is either that life arose only once and all living things stem from this one event; or that the existing genetic alphabet proved to be more efficient than the others and is the only one which endured.' Dobzhansky indeed sees design both in the initial conditions and in the course of evolution. He says,[20] 'Christianity is a religion that is implicitly evolutionistic in that it believes history to be meaningful: its current flows from the creation, through progressive revelation of God to man, from man to Christ and from Christ to the kingdom of God. St Augustine expressed this evolutionistic philosophy most clearly.' It is for this kind of reason that I have argued elsewhere that 'amongst the world religions only Christianity can meet the demands of the scientific world view cognisant of evolutionary biology, and only Christianity can serve a mankind fully conscious of its past and of the evolutionary possibilities of its future'. We shall return to this subject in a later chapter.

NATURAL SELECTION AND ITS LIMITATIONS

About a hundred years ago the great controversy commenced which set against the doctrine of special creation the idea that creation proceeds by an evolutionary process. This evolutionary process was seen as being based on natural selection, a device which was paraphrased by Huxley as 'the survival of the fittest'. And it can be said without hesitation that the general result of the advances made by evolutionary biology in the last fifty years or more have amounted to a tremendous vindication of the theory of natural selection. But though this is true, it is also clear that natural selection has its limitations; and it is important to consider exactly what these limitations are and what their significance is for the view of it as having superseded creation. It is of course generally accepted that mutations are the raw material upon which natural selection has to act in order to lead to what we regard as evolutionary advance; yet these mutations appear to be random and not in any way specifically adapted to the needs – as far as we can see them – of the organism or stock. Then, all that is known about mutation rates, whether in man or plants or animals, tends to confirm the conclusion that they are too low for an organism to be forced by the occurrence of occasional mutations in a given direction; forced, that is, along a line of evolution *against the action of natural selection*. That is to say, there is no room now for the kind of *évolution créatrice* envisaged by Bergson, according to which it was supposed that an *élan vital* could be directly controlling the course of evolution solely by controlling the nature of hereditable variations. In that special sense, belief in the creativeness of the evolutionary process is now a view very difficult to hold. However, it is, I think, true beyond any doubt that *On the Origin of Species by means of Natural Selection* has in the

13

century of its existence had an effect perhaps more profound than that of any other scientific theory;[21] an effect on man's general way of thinking about the universe in which he lives and about his own nature.

The random nature of the variations upon which natural selection has to act has always been one of the biggest stumbling blocks in the path of those who otherwise felt ready, indeed eager, to accept the theory. It has never been put better than by Darwin himself when he said in one of his letters, 'I may say that the impossibility of conceiving that this grand and wonderous universe with our conscious selves arose through chance seems to me the chief argument for the existence of God; but whether this is an argument of real value I have never been able to decide. . . . The safest conclusion seems to be that the whole subject is beyond the scope of man's intellect.' This implies, I think, that Darwin had at the back of his mind this difficulty: If the very idea of mechanism implies an end or goal, how has a finalistic or directive quality been injected into the purely physical universe supposed to have existed before the coming of living things? This puzzle was emphasized rather than reduced by the development, in the twenties of this century, of mathematical theories of evolution by Haldane, Fisher and Sewell-Wright who reinstated the concept of natural selection in its position of supremacy by arguing that even if mutations were expressing a tendency to develop in a particular direction, they were nevertheless far too few to account for any form of evolutionary progress. But in this very act of rescuing natural selection, they reduced it in effect to a truism by showing that all that the term 'survival of the fittest' could mean was that those individuals that leave most offspring will make the greatest hereditary contribution to the next generation.

In the very simplest of organisms, such as viruses, mutations may appear fully 'random', though in fact they are not.[31] As more and more elaborate organisms are developed the very process of natural selection will itself result in an adaptation of the genes which govern the hereditary process, of such a kind that the genes become grouped together in integrated co-adapted complexes. In so far as this is true, it thus becomes that much easier to understand those rather numerous cases of evolutionary development in which we see, or seem to see, the necessity for the simultaneous occurrence of many different and apparently independent variations, each fitting together to produce, for example, special sense organs (such as the eye or the ear) of ever-increasing precision and complexity. The thought of the evolution of the vertebrate eye gave Darwin sleepless nights; but we today find that once any kind of sense organ, whether visual or auditory, has been produced and the genes concerned therewith co-adapted, every improvement however slight will tend to increase in at least some respects the overall efficiency of the organ and so itself allow natural selection to operate according to the original concept of its action.

But even so, there are many difficulties; and many modern biologists, like Darwin himself, may feel the same 'impossibility of conceiving' that natural selection is the sole mechanism at work. This difficulty is at the base of various forms of Lamarckism which have been proposed from time to time, according to which when an animal is faced with new necessities – that is to say, with new needs in the carrying on of its life – it will develop new structures for performing what is required of it. Further Lamarckists suppose that these new faculties will be passed on to the offspring through heredity so that they would result in a true evolutionary change. In fact the pure Lamarckian conception has never been vindicated. But there is a form of it – or rather what we might call a neo-Lamarckian theory – which is highly cogent. Firstly, there is a continuous living interchange between the organism and its environment; and before an organism's environment can exert natural selection on it, the organism must select the environment to live in. That is, there is a feedback or cybernetic system in which there is nothing that is simply cause or simply effect. It is useless for melanistic moths in industrial areas to become darker unless they choose the dark patches to sit on, which in fact they do. Having made this choice, natural selection can operate still further. In other words, with each new organ or change of structure, a corresponding change of behaviour is likely to be necessary if the new development is to be a going concern.

The older discussions of the Lamarckian problem of the in-heritance of acquired characters usually[22] missed the point that all characters of all organisms are to some extent acquired in that the environment has played some part, if only permissive, in their formation. Similarly, all characters are to some extent inherited since an organism cannot form any structure for which it does not have the hereditary potentiality. Whenever a population of a species has been tested for the ability of its members to acquire characters during their lifetime under the influence of abnormal environments, it is found that different individuals differ in their hereditary potentialities for this process. Thus when the larvae of the fruitfly *Drosophila* are grown on a medium which contains sufficient salt to kill a large number of them, the survivors tend to become slightly modified during their development. This modification takes the form of enlargement of two papillae at the hind end of the body which are believed to help in regulating the salt content of the body fluid. Since these larvae are the only ones that survive, there is a strong natural selection for the ability to survive on salt. The end result of this process is that the capacity to acquire the character (enlarged papillae) has been improved; and we finish up with a population every individual of which has a hereditary constitution that makes it very efficient at developing large anal papillae if subjected to the stress of a high content of salt in the food. In other words, the acquired character – namely the enlarged anal papillae – has to this extent become inherited – that is to say, it no longer disappears completely when we remove the particular environmental stress which initially brought it into being.

Another recent development in evolutionary theory has been the recognition, as the result of the work of Thoday and others, of a kind of selection which has been called disruptive or diversifying. Experiments consisted in selecting for twenty-five generations *Drosophila* that had on the one hand a larger, and on the other a smaller, number of so-called sterno-pleural bristles (that is, hairs on a certain part of the fly's thorax). In order to do this, four 'lines' or 'families' were estabished, 'a', 'b', 'c', and 'd'. In every generation all parents of lines 'a' and 'b' had the highest numbers of bristles in the population they came from, and all parents of lines 'c' and 'd' had the lowest in their population. The lines 'a' and 'b' are subsequently called the 'high lines' and 'c' and 'd' the 'low lines'. In every generation the female parents came from the populations they contributed to; males always came from a different population. Thus females having the largest number of bristles were selected from 'a' and 'b' populations; the high 'a' females were crossed in alternate generations to males from the 'b' and 'd' lines likewise having the largest number of bristles in their lines. The high 'b' females were crossed in alternate generations to high males from the 'c' and 'a' lines. In the lines 'c' and 'd', females were selected with the lowest number of bristles. The low 'c' females were crossed in alternate generations to 'd' and 'b' males having the lowest number of bristles, and the low 'd' females were crossed (also in alternate generations) to 'a' and 'c' low males. Lines 'a' and 'b' are accordingly being selected for a high and 'c' and 'd' for a low number of bristles. 'a' and 'b' receive, however, an inflow of genes from the low lines 'c' and 'd' and, in their turn, 'c' and 'd' from the high lines, 'a' and 'b'. The process is shown in Fig. 1. As Dobzhansky points out, a somewhat similar situation is easily conceivable in man. Suppose that one class of people consists mostly of tall persons and another of short, and tall stature is preferred or favoured in the first class, and short stature in the second. It will follow that social mobility in such a society might take the form of the tallest individuals of the short class marrying into the tall class and the shortest persons of the tall class marrying into the short class. The effect of this would be that even social mobility of this sort may not

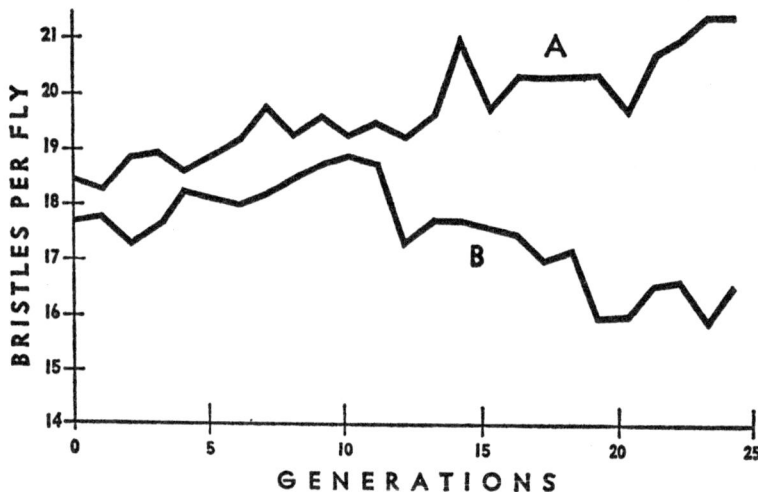

FIGURE I Diversifying selection for the number of certain bristles on the body of the fly, *Drosophila melanogaster*. In the 'high' line A, individuals with the highest number of bristles from the 'high' line were mated in every generation to individuals with the highest number of bristles from the 'low' line. In the 'low' line B, individuals with fewest bristles from the 'low' line were mated in every generation to individuals with fewest bristles from the 'high' line. The experiment showed that the 'high' line became gradually more and more distinct from the 'low' line in the number of bristles.

(from Dobzhansky, T., *Mankind Evolving*, 1962, p. 249)

prevent the tall class getting taller or the short class getting shorter. The figure shows that though the observed course of selection progress appears rather erratic, it is quite clear that the high lines are getting higher and the low ones lower in the number of bristles.

A result such as this could be interpreted as providing support for a conclusion which has often been put forward (e.g. by Professor Darlington) that the existing stratification of the social classes reflects their native abilities. This kind of neo-Nazism, if one may call it that, has I believe been effectively countered[23] by demonstrating that *the quality most consistently favoured in man by natural selection has been 'educability'*. Capacity to profit by experience, to adjust one's behaviour to the requirements and expectations of one's surroundings; trainability for whatever occupations or professions the society has available. They argue, in short, that educability confers the highest Darwinian fitness on human genotypes, that culture is man's most effective means of adaptation to his environment, and that genetically conditioned educability is his most potent biological adaptation to his culture.

Examples such as these, all go to suggest that once some degree of social life has been achieved and once some elements of choice begin to play their part in the life of the organism, the process of natural selection takes on a rather different aspect. Trotter in a famous book[24] assumed that the restrictive influence of natural selection can be mitigated once social groupings have become established. The explanation for this, so he argued, lies in the fundamental biological meaning of gregariousness; which is that gregariousness allows of an indefinite enlargement of the unit upon which the undifferentiated influence of natural selection is able to act. The result is, according to Trotter, that the individual, merged in the larger unit, is shielded from the immediate effects of natural selection and is directly exposed only to the special form of selection which obtains within the new social unit.[25] There is undoubtedly much truth in this; and many prophets of eugenics, basing their views upon it, have vigorously warned mankind of the dangers of the supposed suspension of natural selection among civilized men. They have assumed, with plausibility, that the intelligence and the high level of health and efficiency characteristic of the primitive peoples of the world are being gradually dissipated to leave us with a stock feebler and less satisfactory in almost every respect. There are many arguments to show that at least the worst of these predictions are not being realized. It is argued that the changes which occur in the human condition in historical times have taken place not because the populations were altered genetically but because they were altered culturally. The human species is biologically an extraordinary success precisely because its culture can change so much faster than its gene pool. This is the reason, so Dobzhansky argues, why cultural evolution has become adaptively the most potent extension of biological evolution. He thus supports the attitude of Trotter in its main essentials. But however plastic man as a social animal has become, there are still many examples in the bodily development of man (and not of man only) where the physiological potentialities, and on top of that the mental potentialities, seem to have run far ahead of any conceivable needs or any conceivable controlling influence of natural selection. One of the most striking examples of this is the extraordinary development in size and complication of the human brain, which equip it with aptitudes and facilities far in excess of what (on any even remotely plausible view) can have been needed at any given point in its evolution. There seems in human evolution to have been a trend, a kind of pre-adaptation, proceeding over long periods, irrespective of any foreseeable needs of the stock. Such examples of pre-adaptation, whether in birds, mammals or man himself, often seem to defy explanation. Are they outside the scope of Natural Selection?

Before considering further the question of the creativeness of evolution, we must say something about the idea of emergence, since this is central to the subject and exhibits a chequered history of alternate acceptance as axiomatic and vigorous vilification. Thus to many the idea seems a doctrine of despair. To certain types of mind it suggests some special unanalysable factor which is outside the purview of natural science. To such the proposition that at some stages in the natural process a completely new and in principle unpredictable quality may appear, has an inhibiting effect. But in my view such conclusions should provide not an inhibition but a challenge to further investigation.

The idea of emergence, in its modern form, originated with Lloyd Morgan, Alexander and Smuts. But it early fell into grave difficulties as when organic chemistry, busily fabricating new compounds, successfully predicted their secondary qualities (such as colour) which, according to the doctrine, should have been unpredictable. However, some new and very cogent expressions of the significance and implications of emergence have been produced by Polanyi.

Polanyi commences by reminding us of Laplace's vision of universal knowledge. Laplace argued that if at any moment we knew the positions and velocities of all particles of matter, and the forces acting between them, we could compute the positions and velocities of the same particles at any future or past moment. Thus a knowledge of all things to come and all things past would be available to us. True, today, this mechanical conception of the universe would have to be transposed into quantum mechanical terms, but it would still recognize only one single level of existence, acknowledging no comprehensive entities nor the ensuing stratification of existence. Polanyi proceeds to consider the questions raised by such a statement by referring to the case of machines. He points out that the principles according to·which a machine works cannot be accounted for in terms of physics and chemistry, yet the machine is an inanimate body. How then can it be that physics and chemistry should fail to describe it fully? And if there do not exist superior principles which control its comprehensive action, how can these fail to interfere with the laws of physics and chemistry which apply to the parts? How can it be that the machine actually relies on the laws of physics and chemistry for performing its functions as a machine? Polanyi finds the answer in the view that the laws of physics and chemistry do not determine the *configuration* (my italics) of positions and velocities in which they start to operate. Laplace himself says that the initial conditions have to be given before the physicist can make any predictions. That is to say, any mechanical system can be *shaped initially* according to principles which are not accounted for by physico-chemical sciences; and it may then continue to function in accordance with the same principles while relying on the laws of physics and chemistry. Thus for Polanyi a machine comprises two levels of existence because its initial parameters are controlled by the laws not of physics and chemistry but of technology which cannot be accounted for by the latter. The deficiency of the Laplacian conception faces us more generally in the fact that questions in which we are interested arise in the context of experiences which do not consist in atomic configurations, and which may not be derivable from this conceptual framework. Thus a machine can be described as a particular configuration of solids. The description would state the materials and shapes of the parts and the boundary conditions by which they are joined together as a system. 'But this could describe only one particular specimen of one kind of machine. It could not characterize a class of machines of the same kind, which would include specimens of different sizes, often with different materials, and with an infinite range of other variations. Such a class would be truly characterized by the operational principles of the machine, including the principles of its structure. It is by these principles, when laid down in the claims of a patent, that all possible realizations of the same machine are legally covered; a class of machines is defined by its operational principles.' Suppose, for the sake of argument, that the difficulties of deriving the laws of physics and chemistry from a Laplacian knowledge of the world's atomic configuration can be overcome. We may observe then (i) that a particular specimen of a machine is characterized by the nature of its materials, by the shape of its parts and their mutual arrangement which can be defined by the boundary conditions of the system; and (ii) that the laws of physics and chemistry are equally valid for all solids, whatever their materials and shapes and whatever the boundary conditions determining their arrangement. Thus it follows that neither the materials nor the shapes of the solids forming part of a particular machine, nor their arrangement, can be derived from physics and chemistry. Therefore physics and chemistry cannot account for the existence of the machine. On the bare hypothesis of physics and chemistry you cannot even identify a machine as a machine; and still less identify its working and account for that.

Polanyi points out that in so far as the living body functions as a machine the same conclusions can be readily applied to it also. Physiology consists of operational principles relying on the laws of physics and chemistry which control the parts in which these principles are embodied. Physiology, he says, cannot therefore be accounted for by the laws of physics and chemistry, any more than the operational principles of a machine can be so accounted for. The operational principles of living beings are embodied in the parameters left undetermined by physics and chemistry – in the same way as machines. That is to say, machine-like function, whether in machines designed by human beings or in the machine-like structure seen in living organisms, are based on what can be called the principle of marginal control. Borderline conditions are left open by operations of a lower level (in this case physics and chemistry), enabling the higher level (in the case of human machines, the aims of the designer) to control these borderline conditions which are left open by the operations of inanimate matter. Engineering provides a determination of such borderline conditions, and this is how an inanimate system can be the subject of dual control at two levels: the operations of the upper level are artificially embodied in the lower level which is relied on to obey the laws of inanimate nature.[26]

We are so used to identifying the mechanistic explanation of living beings with an explanation in terms of physics and chemistry that is not easy for many of us to see that could be mistaken. We have to recognize that the living being, even when represented as machine, comprises at least two levels of existence in which the higher may rely on the lower without interfering with the laws governing the lower.

The bearing of these conclusions on the argument from design is far reaching. The only way to reconcile a mechanistic conception of the universe with the fact that it has given rise to the evolution of living beings, is to assume that these comprehensive entities were pre-formed by a suitable pattern of parameters within the mass of primordial incandescent gases. That is to say, instead of rushing about at random the particles of the gas must have been ordered by such a pattern of positions and velocities as would manifest itself as the gas cooled down, by producing living beings and the whole evolutionary development, including of course, man and all his works. But as Polanyi points out, 'the assumption of such an infinitely sophisticated original gas would simulate the comprehensiveness of mechanics only by abandoning the randomness of thermal motions on which thermodynamics is based. And even so, it would be useless. It could explain machines and living beings working as machine but no ordered pattern of primeval gases could account for the sentience of living beings since physics and chemistry know nothing of sentience in matter.

It is clear from these arguments that mechanics must be regarded as a logical emergent from physics, for it is obvious that our interpretation of a mechanism – indeed, our mere recognition of it – does involve something inexplicable on purely physical and chemical assumptions. But does it involve more? It might imply that all living mechanisms are emergents from the physical sciences in the sense that the laws governing living systems are *in essence* different from those governing non-living systems. This may contain an important element of truth; but the evolutionary biologist would, at least for the time being, strongly oppose it for the following reasons: (i) Because of the narrow gap between 'non-living' and 'living', it now seems unreasonable to bet on there being a real and fundamental distinction. The reasons for this I have already discussed. But this after all is a doubtful argument. If fundamental distinctions are to be made somewhere, why not here? (ii) While the evolutionary biologist might agree that no purpose can be discerned in the physical universe prior to the state at which evolution in the biological sense commenced (that is to say, where entities which are born, reproduce and die and in so doing are subject to natural selection) *yet* he might argue that evolution by natural selection automatically provides the 'purpose'. That is to say, he might argue that evolution by natural selection automatically injects something into the cosmos which appears to us as purpose. In other words, once you have a selective mechanism which ensures that forms which tend to last longer become more numerous, then you have the directiveness which is characteristic of biological and ultimately of man-made mechanisms. Thus it is meaningless to ask the question: What is a physical system such as a nebula, an atom or a solar system for? On the contrary, it is always meaningful to ask of a mechanism, whether a biological mechanism or a man-made mechanism, 'What is this for?' [27] (iii) The evolutionary biologist might cite the view that organisms are [28] open systems which display equifinality. By this is meant that organisms are open systems which (exchanging materials from the environment) attain a steady state, which is then independent of the initial conditions. By contrast to this, physical (inorganic) systems are closed systems in which the final state does depend on the initial conditions. As an instance of this it might be argued that the radioactive decay of certain types of atoms is very similar to the mutations and other changes which take place in organisms. The difference is that the decay of the former is independent of the environment – they are in fact closed systems – whereas the changes of the latter are influenced by the environment, i.e. they are open systems. 'The directiveness which is so characteristic of life processes that it was considered the very essence of life, explicable only in vitalistic terms, is a necessary result of the peculiar system-state of living organisms, namely, that they are open systems.' [29] Perhaps this is in fact exactly argument (ii) of the evolutionary biologist that I have just been describing.

Attempts have been made to get over the difficulty of emergence by suggesting that when certain arrangements of the atoms of carbon, nitrogen, hydrogen, oxygen, etc., exhibit the properties which we recognize by the names of enzymes; when other still more complicated arrangements turn out to be able to duplicate themselves identically like the genes in the cell nucleus, or to be able to conduct electrical impulses like nerve cells or even

to exhibit correlated electrical phenomena found in the stagger-
ingly complex systems of nervous cells in the brain; it is quite
unjustifiable to suggest that we have to add something of a non-
mechanistic kind to an already fully comprehended material
atom. What we have done is simply to discover something about
atoms that we did not know before. Referring to this Waddington[30] says, 'there is nothing *philosophically* mysterious about this.
But still it would be frivolous to consider it unimpressive.' Here
he is, I think, in fact begging the question, for where he used the
word 'impressive' he really means nothing less than 'mysterious'.
He goes on to argue that the ultimate constituents of matter –
atoms, electrons and so on – hardly become known to us except
when they combine to form structured entities which have a
definite character; and every time they do so, he argues, they betray
to us a little more of their secrets. But the essential problem which
is glossed over here may be stated as follows: How does an atom
or concourse of atoms know about other atoms and their aggregates? Waddington's answer to this is, in effect, that as archi-
tecture or structural arrangement becomes more complicated, the
ultimate constituents are able to *reveal* more and more of their
own character within it. This finally shows the weakness of the
argument, for what he is here doing is saying that in effect the
fundamental particles have a 'mental component' which can only
reveal itself when they are combined in certain complex patterns
and arrangements. Thus the gases of the primordial nebula were
not just gases at all. Not only did they contain within themselves
the psychic components which, appropriately arranged, would
ultimately give rise to the genius and mental life of a Leonardo da
Vinci or a Bach; the arrangement of these very particles was itself
so far from random that they were compelled to give rise to just
this arrangement and order. As we have seen, to say this is, in fact,
to say that the physical picture of the random movements of a gas
is completely irrelevant; and that really, entirely contrary to
the scheme of the physical science, a stupendous degree of organization was there all the time.

ADDENDUM

Three recently published books bear closely on some of the topics discussed. [31, 32 & 33] On pp. 14–15 above I question the randomness of mutation, expressing doubt which has been an undercurrent of thought in the minds of scores, perhaps hundreds, of biologists for 25 years. Whyte's book crystallizes it. He sums it up thus: 'The conditions of biological organization restrict . . . the possible avenues of evolutionary change from a given starting point. The nature of life limits its variation and is one factor directing phylogeny.' Thus the mutations whose consequences reach the Darwinian test have already been sifted by an internal selection process. This 'internal selection' restricts the directions of evolutionary change by internal organizational factors. He says: *'There may be no mutations which can be fully ascribed to chance.'* He quotes Bachelard (1953) that there is implicit in quantum mechanics a *'structuring tendency'* for complex systems to form more complex forms of ordering. This could perhaps be the basis of internal selection (but see [26] p. 162 below).

Harris[32] quotes (p. 232) T. H. Morgan and L. T. Hogben that selection (both internal and Darwinian) produces trends by canalizing and restricting life to certain combinations from among the infinity theoretically possible. Morgan and Hogben were right in saying that 'if no selection had occurred, all the known forms of life would still have appeared, as well as an enormous number of others.' (But this is a theoretical point only since an infinitely long time would be required.) Quastler[33] says (p. 16) 'The "accidental choice remembered" is a mechanism of *creating* information and very different in nature from mechanisms of *discovering* information.' This can be paraphrased by saying 'selection produces information'. Thus the penetrating dictum of the composer Pierre Boulez (see p. viii above) that artistic creation makes the unpredictable inevitable, shows a resemblance between the processes of artistic and evolutionary creation. Thus, given a finite time, natural selection is truly creative in the same manner as is the activity of the artist. Both can be said to make the unpredictable inevitable.

CHAPTER 2

Evolution, Information Theory and the Emergence of Novelty

DIRECTIVENESS AND ITS ORIGIN

In the previous chapter we were discussing the nature of directiveness in non-living systems, such as man-made machines; and had considered the plausible contention of the evolutionist, that directiveness had been, so to speak, injected into the physical world by the process of natural selection. In this way, so it was suggested, functionally adaptive living machinery was produced; and this eventually led to higher beings, such as ourselves, capable of designing purely physical machines, the directiveness of which is derived from the purpose of the designer.

But this supposed argument of the evolutionist really gets us nowhere; for one must always suppose that the entity showing directiveness had first to appear before it could be selected. Therefore its new directive quality was emergent in the first place. It may be countered that the evolutionist argument *is* sound in that natural selection is the full and sufficient method for ensuring that *accidental* occurrences which have potential usefulness are secured and established for the future biological development of the cosmos. The error here is to ignore evidence as to the low proportions of substance involved. We shall consider below (p. 37) the *extreme* improbability of certain essential steps (of which a number would be required) ever occurring at all. Thus when considering the possible evolution of proteins we have to assume that of 10^{650} possible proteins not more than one could come into being.

28

In the last chapter I described how the concept of borderline conditions, determined in the case of a humanly designed machine, the configurations whereby a physical object or series of objects take on the characteristics of a machine – characteristics which are not specified by the laws of physics and chemistry as they exist at present. That is to say the higher level concepts of engineering, control the borderline conditions left open by the operations of the laws of physics and chemistry. Thus we could conceive a hierarchy in which a higher level can come into existence only by a process not manifest in the lower level.[1] In all these arguments we have to be strictly on our guard against the error which was referred to in the last chapter as 'pseudo-substitution', which connotes to the practice of surreptitiously stretching the meaning of terms in a more basic science to include ideas appropriate only to a 'higher' science. The expressions 'more basic' and 'higher' used here are, I think, equivalent to the terms 'restrictive' as applied to the physical sciences and 'non-restrictive' as applied to the biological sciences.[2] Thus in the hierarchies of systems or sciences by which we attempt to bring order into our experience of the external world, we find in the non-restrictive sciences ideas and concepts which are absolutely essential – in the sense that without them the sciences in question could never have come into being – but which nevertheless find no place and are quite meaningless in the restrictive sciences. The appearance of these new ideas or concepts is, therefore, an example of the appearance of emergence.

What appear to be emergents are appearing constantly around us – not merely as instanced by the secondary qualities, such as colour, of which the holist philosophers spoke, but in the everyday development of organisms. There are two aspects of this: (i) the question of there being a complete and strict physical causality governing every stage of development of an organism with emergent properties, and (ii) the information content of germ cells, of gametes and zygotes, as compared with that of embryos and adults. I am not going to attempt to say any more

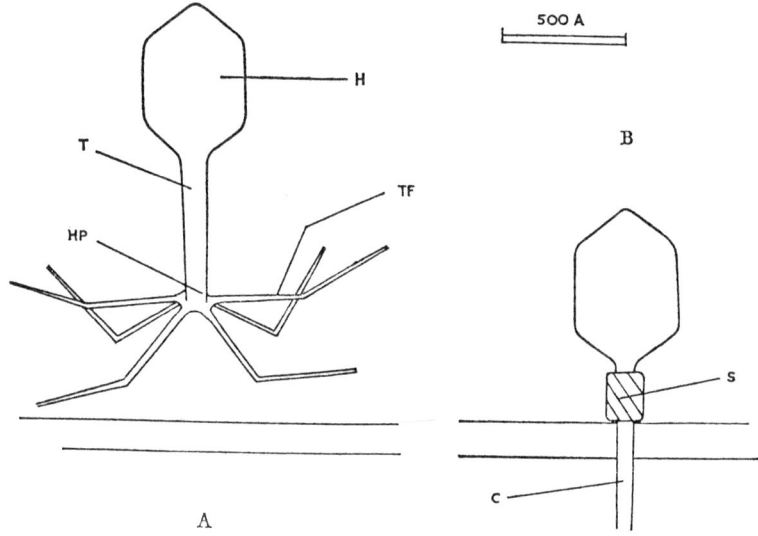

FIGURE 2 Diagram displaying the structure of T2L bacteriophage, based on electron micrographs by Brenner *et al*, 1959. In A contact with the surface of a bacterium is about to be established; in B the sheath has shortened, and the core has passed through the bacterial wall. C, core; H, head; HP, hexagonal plate; S, sheath; T, tail, TF, tail fibre (drawn by M. J. Wells).

(from Picken, L. E. R., *The Cell and the Organism*, 1961)

about causality in modern physics;[3] but the second problem, this question of the information content, is at present a live and challenging issue. What is the nature of the physico-chemical control of form in the development of organisms? The new understanding of gene reduplication has been an impressive step forward. For it suggests the possibility that the modern understanding of the sequences of the four bases of nucleic acids as a 'code' for the production of the twenty amino acids of which all proteins are constructed may be paving the way for a completely mechanistic account of the process of protein synthesis. But great as has been recent advance here, there seems yet to be no plausible or convincing theory as to how the code is 'interpreted' by the cell to enable it to manufacture the required protein. Nor does it account as yet for the non-nuclear element in the hereditary mechanism – which may be very large indeed.

The story is as yet far more speculation than experiment. Facts stick out as widely scattered islands in an ocean of conjecture. Thus Crick,[4] writing on protein synthesis says 'It is remarkable that one can formulate principles such as the sequence hypothesis, and the central dogma, which explain many striking facts and yet for which proof is completely lacking. This gap between theory and experiment is a great stimulus to the imagination.'

But there are some convincing facts as to the molecular determination of animal form – that is to say, of animal mechanism – which seem, at any rate superficially, to argue on the other side – namely that purely physical systems may be directive and that a strictly physio-chemical system may sometimes define the form of an organ, as it does of a crystal. Fig. 2 illustrates the T2L bacteriophage as seen in electron microscope photographs close to the bacterial cell wall. Each part of the structure is assembled spontaneously from distinct chemical sub-units because it is of the nature of the sub-units to behave in this way. Picken[5] says, 'There is no discontinuity between the molecular and ultramicroscopic structural levels; the sub-microscopic shape is an expression of combining properties of large molecules in a specific environment of smaller molecules and ions.' Fig. 3 illustrates the sperm cell of the crustacean *Galathea*. Although the size range is different, we see an extraordinary superficial resemblance and here again the structure can be very largely if not entirely 'explained' as a direct result of the chemical constitution. The same is true to a considerable extent of some of the recent studies of the sperm of mammals. Both of those illustrated are in fact armour-piercing bombs each of which injects nucleic acid into its objective; in the first case resulting in destruction, in the second resulting in development. Here then are two structures which appear to be mechanisms, in any ordinary sense of the word, in

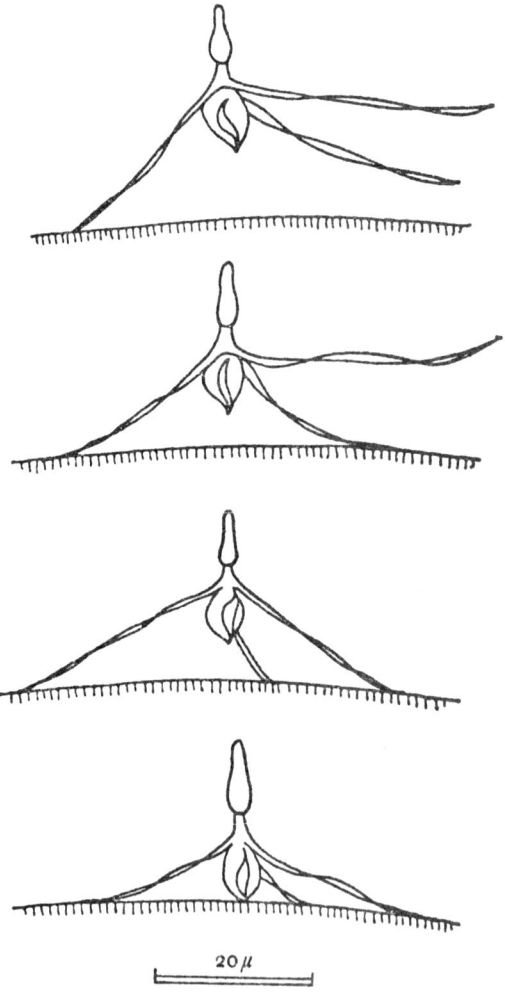

FIGURE 3 Four stages in the attachment and orientation of a sperm of *Galathea* at the egg surface. The progressive extension of the area of adhesive contact between the three processes and the egg surface brings the axis of the body of the sperm normal to the surface and insures that the body is in immediate contact with that surface (drawn by M. J. Wells after Koltzoff, 1906).

(from Picken, L. E. R., *The Cell and the Organism*, 1961)

that they appear structurally adapted to perform a particular function. These are certainly objects of which we can sensibly ask concerning the whole or a part, 'What is this for?' Yet they are also objects which can be at least plausibly explained solely as the direct outcome of their chemical composition. Yet, while admitting this, one must also point out that the full significance of the form can only be grasped when we bring in the biological concepts of function and adaptation. But we can at least say that this is one of those innumerable 'mechanisms' found in the world of living things for which natural selection alone does not seem a reasonably adequate explanation.

Examples like this, and probably we shall soon have a comparable chemical specification of many others, suggest that on the face of it the distinction between the non-directiveness of chemical systems and the directiveness of mechanisms may not always be meaningful. We shall certainly know much more about this kind of situation in the years immediately ahead.

But to come to this matter of information content. What light may estimates of the information content throw on the matter?[6] Information theory arose as developments in communication led to the realization that general properties related to what may be called 'the assembly of a pattern' exist. Thus we can regard a picture on a television screen as the expression of a definite number of all-or-none signals in a channel – that is, the number of signals necessary to specify that particular picture. Setlow and Pollard[7] have defined information as follows: 'if an event has a probability p of occurring before a "message" is received and a probability p' of occurring after the message is received, the information in the message is H where $H = \log_2 \dfrac{p'}{p}$. To see how this operates, suppose that we have to pick out one letter from a total of sixteen. Then, before picking, $p = 1/16$, while after selection is made, $p' = 1$, so $\dfrac{p'}{p} = 16$. Now $2^4 = 16$ and $\log_2 16 = 4$, i.e. there are four bits of information.' The H in the above formula is in what are called 'bits', this being a contraction of the term 'binary units'. This is the simplest kind of definition of information. If all the possible states are equally possible, but the expected value of the i th state is p_i, then the definition of information follows this formula: $H = - p_i \log_2 p_i$. So we can see that the information content of a complex pattern can be expressed numerically as a certain number of bits; whether the information is in the form of a picture or a sound pattern; whether it is in process of being transmitted along a channel; or is safely stored (as it is presumed to be stored) in the DNA or other macromolecule of the chromosome, or in a cellular or sub-cellular structure of the brain.

The task of estimating the information content of biological systems such as a virus particle or a bacterial cell is relatively simple and can therefore be relatively accurate. The task of estimating the information content of a higher metazoan is of formidable difficulty and, as we shall see, only the roughest approximations (usually little more than plausible guesses) can be arrived at. Nevertheless, even here the approach of information theory can be most valuable, for it enables us to gain *some* idea of how complex a system really is; and it may give us a clue as to whether a given hypothesis for a particular biological process or event is reasonable or totally unreasonable. As Setlow and Pollard[8] say, 'it does not tell us how to do something, but rather *how difficult it is to do*'. Several estimates have been made of the information content of a bacterial cell. Two of these, one by Morowitz and the other by Linschitz,[9] are worth considering. The first is in terms of the direct approach of information theory, the second involves estimates of physical entropy and its relation to information. A bacterial cell is of course made of water and solid material; whereas a bacterial spore lacks nearly all the water yet can, nevertheless, develop into a bacterium. So one can assume that the information content is primarily in the dry part; and the problem then is to choose the right atoms and put them in the right places. The instructions for doing this in binary form are the information content. There is large scope for error in deciding on the fineness of the grid. If one is preparing instructions for a building made of bricks, then a three-dimensional grid can be imagined forming a three-dimensional honeycomb of cells. One then reduces the amount of information to a number by deciding whether the brick is or is not in a cell. Clearly it will make an enormous difference whether one imagines a coarse or a fine grid. If the grid is too coarsely designed to describe the building, then the resulting specification will be inadequate. If, on the other hand, it is too fine, the description may include slight cracks in the bricks which are not relevant to the design. In spite of this kind of difficulty the Morowitz method works out at approximately 10^{12} bits for specifying a bacterial cell; and it is interesting that Linschitz, working in a very different manner, arrives at a value of $5 \cdot 6 \times 10^{11}$ bits – which, all things considered, is remarkably close to the estimate of Morowitz. It is, of course, quite possible, indeed it may be virtually certain, that many of the actual locations of atoms are not critical to life. And with these and other reservations it is clear that the value of 10^{12} bits is very high.[10] If the walls of a building are constructed by pouring concrete into moulds instead of the brick by brick method, the amount of information needed decreases. Similarly, if in a cell there is a possibility that certain gross patterns are *required* to develop from others – e.g. if a string of nuclear proteins is placed in a mixture of a few enzyme molecules, salts and aminoacids and then a cell *has* to result – the information content is far less. For a bacterium, the absolute minimum estimate grounded on such reasoning gives 10^4 bits.[11]

If the figure of 10^{12} bits per bacterium is used, then 10^9 bits per second are involved in the growth and development of the organism. But 'under no circumstances can the growth of a bacterium be looked upon as producing less than 1,000 bits per second', and it is significant that J. W. L. Beament (pers. comm.) has estimated that the information content of a bacterium is greater, by several orders of magnitude, than that of the nucleic acid which it contains. And there are reasons for suspecting that information storage and replication may also be occurring in the cell cortex of the amphibian zygote.[12]

To get these figures into perspective, let us compare them with one or two other estimates. Conscious handling of information by a human being works out at approximately 25 bits per second. It has been estimated [13] on the basis of channel capacity necessary to cope with the '*hi-fi*' reproduction of a forty-five minute symphony, that such a work is equivalent to an information content of about 4×10^7 bits and will equal a transmission rate of 15,000 bits per second. The information content of a large picture which is to be transmitted in colour on to a television screen so as to be regarded by an expert viewer as perfect would again involve about 10^7 bits. A similar estimate for the information content of a human body is of the order of 10^8 or 10^9 bits; and for the lifetime accumulation of information in a human brain between 10^{15} and 10^{20} bits.[14] To relate these to certain physical estimates, it is thought that the total number of protons in the universe is about 10^{80} and the total number of seconds which have elapsed during the existence of our galaxy is at the most 10^{18}. Eddington once lightheartedly suggested that a batch of monkeys strumming on typewriters would, if they went on long enough, eventually produce all the sonnets of Shakespeare. Elsasser points out that even if we are a little more modest and consider only the beginning lines of each sonnet and assume these to consist of a hundred symbols each, these can be combined in 10^{143} different ways, so that Eddington's proposal was not only lighthearted, but also utterly irrelevant. In fact some of these calculations are a most valuable corrective for some of those who argue lightly about the high probability of the random origin of life in the universe. Thus[15] any theory of the origin of life must take account of the extraordinary biochemical similarity of all living matter. The fact that only one pattern of optical asymmetry is found in life strongly suggests a common origin for all existing living matter – as W. H. Mills pointed out.[16] Thus the event which produced living matter must have been highly improbable even under

36

primordial conditions. Assuming that an aqueous solution of amino acids had been formed, the next problem is how could these be built up into complex proteins or enzymes? Assuming that the concentration of each free amino acid is kept at one M, the equilibrium concentration of a protein with 100 residues (MW about 12,000) is 10^{-99}M which represents 1 molecule in a volume of 10^{50} times the volume of the earth. This appears to rule out the possibility of the formation of any protein by mass action, even in the presence of a catalyst to bring the system into equilibrium. An amount of protein MW 60,000 equal in weight to the earth contains 6×10^{46} molecules. The number of possible proteins in this molecular weight is 10^{650}. Therefore at any instant not more than 1 in 10^{603} possible proteins could exist at all. Thus the simultaneous formation of two or more molecules of any given enzyme purely by chance is fantastically improbable. Quite apart from any further difficulties in the random theory of the origin of life, such as that of holding the components consistently together until a cell membrane is formed, Dixon and Webb[17] conclude by saying, 'thus the whole subject of the origin of enzymes, like the origin of life, which is essentially the same thing, bristles with difficulties'. We may surely say of the advent of enzymes, as Gowland Hopkins said of the origin of life, that it was 'the most improbable and the most significant event in the history of the universe'. Such calculations, though seemingly remote from our present line of inquiry, are in fact highly relevant to it; because it is often useful in general and theoretical discussions, as it is in mathematics, to consider the limiting case. Looking at this kind of problem in another way, we see it as the problem of the origin of order. Let us once again, before coming to biology, go to the cosmic scale (again the limiting case) and see what estimates we can arrive at.

If we consider the process of cosmic evolution from a primordial nebula through the evolution of the solar system allowing, in its turn, organic evolution to take place, and if we look at the process from the point of view of information theory, we can say that the result has been, in the living universe as we know it, a stupendous increase of order – that is to say, a stupendous increase of information. It is no answer to say that since, for all we know, there may have been no overall increase of information in the universe as a whole, therefore the merely local increase of order which we perceive on this planet can be regarded as accidental. The numerical illustration which I have just given, showing the impossibility of assuming that monkeys hitting typewriters could ever write the sonnets of Shakespeare, applies *a fortiori* to the argument that the evolution of life in the solar system is fortuitous.

First, the apparent increase in information in the life of the individual organism when we come to compare the germ cell with the body of the mature individual. Secondly, a separate though allied problem, the undoubted similar increase in information in the nervous system of an animal as a result both of its genetic make-up and of its experience.

Applying the kind of calculation alluded to above in relation to the information in a bacterium, we can say without fear of doubt that even the information content of an amoeba must be several orders that of the information store of the most advanced computer. Further, it is reasonable to assume that the information content of a highly evolved mammal is far greater than that of an amoeba. Are we to suppose that all the information content, of the adult chimpanzee for example, is somehow packed into the nucleus of the fertilized ovum? That would be a new and I think untenable form of preformationism. It is clear that new information arrives from somewhere during development. Where does it come from?[18] [46]

It is one of the fundamental postulates of those who developed information theory that, if information is to be conserved, either in a store or in a channel, then 'noise', in the sense of the disruptive effect of influences such as thermal agitation, must be excluded. Noise which is random, is the enemy of information; and the only known way in which information can be conserved in a noisy channel or a noisy store is by redundancy. Therefore an engineer designing a store or channel attempts to protect it from noise and to ensure that the information present survives unavoidable noise by having sufficient redundancy. As Elsasser[13] has pointed out, the stability of statistically stored information depends on the mechanical and chemical stability of the storage medium. The medium must be a chemically stable solid body of sufficient mechanical strength, and the storage systems of the engineer are designed on this principle. The engineer does not have to bother with information stores of extremely small dimensions, so he can keep the loss of information due to thermal noise at a reasonable level. But what are we to make of the germ cell, which may have to store information quantities of the order of 10^{12} bits or 10^{15} bits in a space of radius less than 2 microns? As Elsasser says, [13] the problem of maintaining information systems of this size becomes more and more elusive and has not in fact been considered by the engineer or physicist. Elsasser, as a physicist, is so impressed by the difficulty that he finds it impossible to believe that the necessary amount of information can be physically stored in the germ cell. He assumes that other laws – biotonic or epigenetic laws, as he calls them – which must be studied independently of physical concepts, govern the development of organisms. He thus thinks that it is the use of tools provided by information theory which will show us the extent to which – if at all – the behaviour of organisms deviates from the behaviour of machines in this vital respect.

Now it must be admitted that if the information stored in the germ cell is exposed to anything like the thermal noise which would be supposed by an engineer, then the conclusion is correct. For under no circumstances could there be room in the germ cell for redundancy of the degree necessary to ensure that no essential information is lost.[18]

An alternative view is taken by many molecular biologists and biophysicists who argue, at least by implication, that the discovery of the double helix of DNA has provided in fact just the mechanism – on an ultra-microscopic scale, yet still resistant to thermal noise – which is needed to salvage the mechano-chemical view of the storage of information in the germ cell. They argue that the resistance to thermal noise of the double helix is tremendous, and that this is because of the binding energy of the bonds in the double helical structure which, being wrapped in upon itself, should be capable of resisting thermal degradation to an almost infinite extent. They sometimes speak as if the nature of the double helix is such that there is no need for redundancy at all, and the problem can be dismissed. It is almost certain that this is an overstatement, but it is extremely difficult to get any precise agreed estimate as to how far it is an overstatement. There is certainly a basic redundancy in a DNA store due to the fact that four bases are used to specify 20 amino acids whereas they would be sufficient, theoretically, to specify 64. In fact there is more redundancy still, since all the 20 will never be present in equal proportions, therefore less information would be sufficient to specify them than is in fact basically available in the DNA system.[19]

For the time being, then, it seems that we must be content with this meagre information. When it comes to the development of the organism, and transcription of the information from the germ cells to the tissues, then all agree that much information must be lost by noise in the transcription process. But here, it is argued, much more can be tolerated because a certain proportion of misfits and imperfections are allowable and can be eliminated by selection, just as deleterious mutations are weeded out in the process of evolution.

The upshot of all this is that in the development of the cosmos, as in the development of an organism, we *appear* to be confronted with a stupendous increase in information. When we come to consider the information increase in ontogeny, that is in the life development of an organism, we find two main aspects, both of which involve the idea of coding.

Until recently I had myself tried to overcome this problem of determination of development by the DNA of the nucleus by assuming that the instructions for development provided by the

DNA and the chromosomes simply mean that, as the organism commences and continues developing, it has within it orders to take certain things from its environment and to leave others; to accept certain kinds of stimuli and to leave others; and so to build up its own internal milieu by taking from the environment at every step the materials and the stimulation that it needs. This is – so I thought – what the genetic instructions are doing. This is the programme to which they insist that the animal conform. So it would follow that at every step in its development the animal is obviously dependent upon its environment; but with each step it becomes increasingly more able to make and control its own environment. On this view, the complexity which an organism displays arises primarily from instructions in the germ cell and not from the instructions which are contributed by the environment. Nevertheless, as the internal milieu of the organism itself is also an extremely complicated environment for a developing organ such as a nervous system, and as this internal milieu is itself the outcome of instructions laid down in the germ plasm, and as there has been interaction between these two throughout the life of the developing organism, so we have a physico-chemical explanation of the origin and increase of complexity during the course of development.

But as a result of the arguments outlined above I am now convinced that this view, which I held until so recently, is inadequate; it can provide a full explanation only in certain special cases, and cannot possibly be generally adequate. If we look at the simplest organisms, such as those plants which can be grown indefinitely in very simple homogeneous media, we find that, in addition to the half-dozen elements which are the basic constituents of the large molecules necessary for life, they do not seem to require more than a dozen or so trace elements. Not only are the elements limited, but the compounds required by such an organism are also strictly limited. That is to say, the food is homogeneous whereas the organism is extremely inhomogeneous. Furthermore, organisms using more complex food must break it down into simple components (e.g. proteins into simple poly-peptides; as is shown by the violent immunological reactions which occur when a foreign protein is directly injected). There-fore we now seem to arrive at the conclusion that even where the hereditary material is inadequate to carry the necessary load of information, the environment cannot be supplying it! This seems to push us to a new form of the old vitalist position which assumed that in living organisms purely physical laws no longer operated; matter in organisms obeyed organic laws which it was impossible to reduce to the terms of physics.

Vitalists of previous generations claimed that various biological functions, such as self-reproduction, self-organization under varying conditions, and learning by trial-and-error, are specific to living things, never to be explained except by a biology which is an entirely separate discipline divorced from physics. It assumed too that these laws were incompatible with the laws of physics; and that therefore sooner or later one would find the physical laws being broken. But latterly we have come to see that such functions do not necessarily constitute examples of independence of physical laws. We now know that automata can be built, or at least conceived, which will perform the equivalent of innumer-able biological functions which the earlier vitalists assumed could never be explained by physics, and so in the last twenty years it has looked as if physics is rapidly preparing to take over more and more of what has previously been regarded as biology; and this is one of the factors which determines the present-day emphasis on a relatively new subject, namely biophysics. Elsasser, and others have emphasized that the fact that automata can do many things that living organisms do, does not in any way prove that living organisms do them in the same way as automata. In fact we have long known that in many instances organisms achieve these re-sults in an utterly different way from that in which a humanly designed machine would proceed. As Quastler[20] has pointed out, the mechanisms for memory in living organisms are very different from anything an electronic engineer would or could design, in that the germ cells are fantastically small considering the enormous amount of information that they store. Again, the memory functions which maintain the stability of adult organisms reside in soft tissues, and not in the form of permanent records maintained in the hard structures, which is where an engineer would place them. Again, years of research have on the whole signally failed to establish the exact place and mechanism of the storage of memories in the brain. There does not seem to be a particular memory organ for particular kinds of memories: it is rather as if all memory, in the higher mammals at any rate, is somehow stored everywhere in the cortex – this is in fact a corollary of what the famous American physiologist Lashley enunciated as the 'law of mass action'.

When the science of embryology was founded, there were those who reported a tiny foetus within the human egg or sperm. They believed that every feature of the adult was already preformed in miniature within the germ cell and needed only to grow and to expand to produce an adult. 'When microscopes were improved, and this view of the matter was overthrown by the actual facts, the concept of preformation was replaced by epigenesis, which holds that the characteristics of the adult are not present in the germ cell but are gradually elicited by the interaction of specific stimuli (either internal or external) with the pre-existing structure of the developing egg. Since then detailed experimental studies of development have proved, at least to the satisfaction of most embryologists, that most developmental processes are epigenetic rather than preformationist in character.'[19] Commoner[21] points out that modern views regarding the genetic function of DNA are in conflict with the concept of epigenesis. If one accepts the conclusion, as does the embryologist Raven[20], that all information in the body of an adult organism, however highly developed, is in fact present in the germ plasm, and that the difference between the germ and the adult is only in the degree of redundancy – if one assumes this, then one is back at a new form of the ancient preformationist position. On such a view one has to believe that in development the pre-existing code is translated, without loss or gain in information content, into the characters of the fully developed adult. There must be in fact a one to one correspondence between the final features and their included representation in the zygote. This has been called 'a modernized version of the preformationist principle'. Quastler[23] states, 'there is a very real homonculus in every sperm and in every ovum; but it does not look at all like a little man any more than a description looks like the object described'. Thus in the language of information theory the DNA code hypothesis holds that the information content of the adult is entirely and wholly derived from the information content of the DNA of the germ cells, while epigenesis means that some other sources of information must arise during the course of development. A major requirement of the DNA hypothesis is the principle that while information which is acquired by the nucleic acid may be recovered from it, information transferred from nucleic acid to protein can never be retrieved from the latter. This is what Crick calls the 'central dogma' and it implies that DNA is a self-duplicating molecule to which protein cannot contribute any information. This concept has been criticized by Commoner[22] and by Dean and Hinshelwood who argue that DNA is not merely self-duplicating; there is still a possibility (some would put it much higher than a possibility) that enzymes and proteins may be contributing information and may be essential to the DNA reproductive process and that if this is so, one simply cannot have a minute scale store without an enormous redundancy, since even if the DNA is resistant to thermal noise, the other components of the duplicating mechanism will not be so.

Dixon and Webb,[24] have some very relevant remarks on the crucial question of information transfer and storage in the intracellular mechanism for protein synthesis. Thus they say:

> The code is built into the biosynthetic mechanism in two places: (*a*) it is embodied in the DNA of the genes, which use it to represent the structure of the corresponding proteins,

44

and (*b*) it is built into the specificity of the ligases. One of the most interesting and fundamental questions in biology is how it comes about that the genes and the ligases use the same code. The answer that no life is possible unless they do is unsatisfying; the chances against it coming about without some controlling mechanism to relate the two are enormous, but it is extremely difficult to picture such a mechanism.

And again:

The structure of both the specific centres in an enzyme which is subject to feedback inhibition are determined by its structural gene. The genes indeed display an astonishing amount of 'knowledge' about the sequence of chemical processes in metabolism. One may well ask how the gene forming enzyme 2.4.2.14 'knows' that phosphoribosyl pyrophosphate will be converted by the consecutive action of ten or more different enzymes into a purine nucleotide (see Table XII, 18), or how the gene for the first enzyme of histidine biosynthesis, which acts on the same compound, 'knows' that its product will be converted into histidine by a different series of enzymes. Even with this information, how do these genes 'know' what aminoacid sequences in their enzymes will act as specific centres combining with purine nucleotides or histidine respectively? Evidently there must be some mechanism whereby information derived from the metabolic processes themselves is transmitted back to the genes and there incorporated in the form of polynucleotide sequences. The manner in which control was established in the first place, and the nature and mode of action of this mechanism, is one of the most fascinating and fundamental questions in biology.

Here the controversy must rest for the time being. Obviously the most immediate need from the point of view of the ethologist is to provide facts upon which better and more reliable estimates of the information content of the adult organism, including its behaviour, can be assessed. The detailed specification of a peacock's tail (to mention a particular example) seems on the face of it to be likely to be straining the resources of any preformationist code in the germ plasm to breaking-point. But what of the details of innate behaviour? How about the coding of an innate bird song? A coding which may govern pitch, time, duration, rhythm and accent, and perhaps timbre as well. Perhaps an even greater problem is set by innate mechanisms for solar and stellar navigation; only further research will reveal how severe this requirement really is. We also certainly need much more data on the relation between quantity of DNA in the germ cell and the complexity both of structure and behaviour in a wide range of animal species. The present meagre evidence does not suggest that there is any close correlation between the two. At any rate it seems to me that the experimental analysis of behaviour may in the end provide data which will be crucial to the solution of this most fundamental problem.[6]

The further arguments on which this is all based are largely outside the scope of the present book. They involve a subtle development of Niels Bohr's theory of generalized complementarity, and a new principle which is enunciated as the principle of finite classes. The argument runs something like this.[25] Consider for example the genetic information carried in the germ cell and presumably represented as a specification of molecular structures. Given the number of molecular building units within a germ cell, the number of possible specifications is immense – larger by several orders of magnitude than the number of existing species. It is therefore possible that the genetic information characteristic of any given species is compatible with a vast variety of molecular specifications of germ cells. If there is actually great variety of molecular configuration, then the elaboration of relatively uniform characteristics from a huge variety would have to be done by biotonic or epigenetic functions. Such a process is in fact beyond detailed analysis; the complete study of the molecular constitution of a given cell would entail more measuring operations than the cell could tolerate without being destroyed; and the derivation of a composite picture from different individuals is impracticable because they are thought to differ at the molecular level and only a minute fraction of all of them could be investigated. Under these circumstances no physical observation could disprove the hypothesis of a biotonic or non-mechanistic information transmission.

To return to Elsasser. He argues that when dealing with systems of a relatively simple structure such as atoms or molecules, one can carry through a formal proof showing that the validity of the laws of quantum mechanics excludes the possibility of other independent laws supervening in the same systems. For systems of immense structural complexity, on the other hand, a corresponding proof of incompatibility would involve the use of immense classes but still nevertheless could be (at any rate in principle) carried out. With organisms we have not these immense classes to deal with. Suppose we try and set an upper limit to the membership of any class by estimating, as Elsasser does, the total number of cells on earth. Making certain plausible assumptions, Elsasser arrives at a total number of somewhat below 10^{28}. Assuming now that, due to metabolism, cells change their molecular configuration rapidly, so that we must count each cell as a new unit every minute of time. To quote Elsasser, 'if organic life exists on the earth from an early geological age until the sun has burnt out its nuclear fuel, say fifteen billion years or 10^{16} minutes, the total number of cell configurations thus defined is 10^{44}. But while this upper limit to the number of members of a class is very large indeed, it is by no means an immense number in the meaning given to the word in this context.' The logarithm 44, while a moderately large number, is not extremely large in the sense of being of the order of millions. The number of members in any class of organisms is thus immensely small compared with the number of microscopic configurations of which the members of the class are capable. Even the supposition that life exists on, say, 10^{20} other planets would not change the order of magnitude relationship. From this it appears that the failure to prove contradiction between physical and biotonic law (which failure must

47

be assumed if we are to maintain the existence of biotonic law) could be due to the actual limitations on the size of classes or organisms. To quote again: 'If it is true that proof of incompatibility requires immense classes, then it is unlikely that enough specimens of any one class can ever be found to demonstrate it.' To give a primitive example, if an urn is filled with a large number of balls of various colours, then the result of drawing just two or three balls does not entitle us to make an inductive inference about the relative proportion of colours among the balls. Similarly, the impossibility of establishing certain inductive inferences in biological theory is due to the limitations of the size of the classes. In other words, no physical disproof of the existence of biotonic laws of organization can ever be given.

From arguments such as these Elsasser enunciates the principle of finite classes as follows: '*Classes of organisms are immensely small as compared to the number of microscopic configurations that the members of the class may assume.* The dynamics of these systems are such that biotonic laws can exist compatible with any possible physical prediction for the class; the membership of a class will be exhausted before a statistical proof of contradiction can be achieved.' Thus we can think of the organism as a dynamic whole relative to which both purely mechanistic and purely biotonic modes of behaviour appear as limiting approximations.

Since this part of the theory was developed, Elsasser has given reason for supposing, or suspecting, the existence of specific devices or processes which are designed so as to create in the organism a high degree of physical impredictability. All organisms are in fact intrinsically inhomogeneous. He quotes two biologists to support him in this, R. D. Hodgkiss who says, 'Life in the repetitive reproduction of ordered heterogeneity; if you have this you can't have two individuals that are exactly alike,' and Paul Weiss who says, 'the idea of identity of cells, any two cells, any two individuals and so on, is a fiction'.

Because of the nature of the case and implications of the principle of finite classes, Elsasser concludes that the properties of living things which are organismic are such that no possible experiment or set of experiments can ever reveal a contradiction between the irregularities to which these properties are subject and the laws of physics[26] (quantum mechanics). In other words, the fragmentary evidence obtainable from small classes, which is all we can ever hope to have, can never provide us with the means of saying at what point laws other than the laws of physics intervene. But Elsasser assumes that if enough examples could be studied, compatibility might be demonstrated. Polanyi[27] takes a more fundamental standpoint in these matters – as might be anticipated from what I have already said. He would argue that the nature of operational laws is such that there can never be compatibility, since in his view it is axiomatic that all the laws governing matter in the configurations it assumes in organisms – that is to say, including the borderline or limiting conditions – can never be deduced from atomic configurations alone. Elsasser himself seems ambiguous on this point, and in places speaks as if he would agree with Polanyi. Thus he argues that the laws of organismic biology do not abrogate the laws of physics but on the other hand cannot be derived from them. He points out that there may have been a time when the universe, or perhaps some part of it, contained only protons and electrons and no complex nuclei. Later on, stars formed, and at the high temperatures prevailing in their interior, complex nuclei were synthesized. He says,[28] 'it does not follow that the laws of nuclear structure and dynamics can be reduced to or derived from the much simpler laws of the motion of protons and electrons'. Similarly laws governing organisms were observable only when certain specialized physico-chemical conditions made their appearance on the earth.

As regards evolution, he would argue that, at least in the present state of our knowledge, evolution theory cannot be used either to support or criticize organismic theory. In this respect Polanyi[29] makes a very important point when he argues that the problem of evolution is much misunderstood, if not misrepresented, as a result of the emphasis of current theory upon the facts of natural selection. He says, 'the interest of evolution lies in the rise of higher beings from lower ones and, principally, in the rise of man. A theory which recognizes only evolutionary changes due to the selective advantage of random mutations cannot acknowledge this problem. All forms that continue to survive have the same survival value; only those that are becoming extinct can be said to lack selective advantage.'

Burnet[30] may be right when he says in his book, 'the attempt to press the structural, physical and chemical approach to the understanding of living processes seems to have reached the phase of diminishing returns for the effort involved. We are approaching an asymptotic barrier and it may be that some modification in the outlook and approach of theoretical biology will soon be needed.' I personally think this is too pessimistic in that it fails to allow for the modification of physics and chemistry by biological principles. It certainly seems to be true that the DNA code cannot be the whole story of the mechanism for inheritance and is perhaps only a very small part of it. Moreover[31] the supposed template self-reproduction mechanism of the DNA molecule is physically of immense improbability. Part of the weakness of the DNA hypothesis results[31] from its failure to consider *all* the principles of inheritance and development. Commoner adds, 'to build a successful bridge between *biology and physics the structure must be equally based on two essential foundations: the basic principles of physics and the basic principles of biology. This task still lies before us.*' The bridge we have is insecure at both ends!

MIND AS EMERGENT

Finally we come to the question of mind as an emergent; and this of course has always been the central problem of those who are developing the theory of emergence as a philosophical system. Here particularly we continually come up against the effect of pseudo-substitution. Many biologists assert, or speak as if asserting, that biology explains living things in terms of physics and chemistry. As Polanyi again points out, the purpose that biology actually pursues and by which it achieves its triumphs, consists of explaining living beings in terms of '*a mechanism founded on the laws of physics and chemistry*, but not determined by them'. To condense the question of the origin of mind almost to the point of absurdity, I would like to make a few points.[32] First there is the epistemological aspect. Mind is more than a calculation or representational mechanism in that it involves self-consciousness. Popper[33] has shown that calculators are incapable of answering all questions because it is impossible for them to possess completely up-to-date initial information about themselves. If self-consciousness is in any sense a fact, then minds have a fundamental quality which distinguishes them from calculating machines; for calculators cannot 'think', MacKay[34] has put it thus. 'It goes without saying that so lively and inquisitive a newcomer among academic disciplines is bound to raise philosophical questions – (to take an example) we have heard people ask "Can computers think?" The answer is not "yes" or "no" but "don't be silly". In the case of human beings we do not say that *brains* think, but that *people* do. Computers (as mechanisms in metal) are logically on the same footing as brains (mechanisms in protoplasm), and to say they think would simply be to drop a philosophical brick. Computers manipulate, not ideas but tokens for ideas (as, presumably, do brains).'

Professor MacKay[35] has recently (1966) taken this argument further in an important discussion concerning determinism and free will. Employing the argument referred to above as being used by Popper about computers, that no information system can embody within it an up-to-date and detailed representation of itself *including that representation*, he points out first that it is the working assumption of neuro-physiologists studying the brains of animals and men, that all conscious human activity and experience – choosing, seeing, believing, etc. – has a correlate in corresponding neural activity. It follows that if all cerebral changes were physically determined by prior physical functions, including other cerebral changes, then the future content of conscious experience would in principle be predictable from these factors. It would seem to follow from this that our conviction of having 'Free Will' is an illusion. Really we are bound to do what we do because the 'choice' has actually been forced upon us by our brains. Professor MacKay believes this to be a mistaken conclusion by virtue of the same argument which Popper used about computers.

The key point is that if what a man believes affects correspondingly the state of his organizing system no complete up-to-date account of that organizing system could be believed by him without being *ipso facto* out of date.

By the same token, even given the most complete current data, no complete prediction of the future state of his organizing system is deducible upon which both agent and observer could correctly agree.

A prediction made *secretly* by a totally detached observer may well be valid for him and his fellow-observers; but upon the agent himself it has no logical binding force. On the contrary, he would be mistaken if he believed it even though the observers were correct to believe it.

This is a curious and extremely important kind of *Logical Relativity Principle*. In order that the observers may validly believe what they do about the agent's brain, it is necessary that the agent should believe something else; for if he believed what they do, it would not be valid for him or for them.

To put it in a different way – *no prediction of a typical choice could be believed by the agent without affecting its own validity.*

If we are right that the understanding of Mind will not come by finding some unexplained details of physical analysis but by a new understanding of the significance of brain-and-body activity – then Mind has a working contact with matter, more intimate than that of one form of energy upon another; for they are truly complementary – the mysterious unity we know as 'personal agency'.

Finally, to stress the main point once again, no amount of physical knowledge of the state of a (brain) mechanism can generate a prediction of a typical action which has binding logical force upon the agent whose brain it is. 'In short, our belief that we are normally free in making our decision, so far from being contradictable, *has no valid alternative* from the standpoint even of pre-Heisenberg physics. In this sense at least, we are irreducibly responsible for the conscious control of action.'

Apart from this it must be pointed out that the laws of physics are inferred from experience, but the existence of thought is immediate experience. Bertrand Russell made I think just this point when he said, 'I hold that whatever is known without inference is mental.' But for me the argument as to self-consciousness which influences me most strongly runs as follows: The fact that consciousness has been evolved is itself strong presumptive evidence that it is not an accidental side product. It suggests that at least this aspect of mental life can accomplish something for which a neuronal mechanism alone, however complex and elaborate, is inadequate. How far down the scale this concept of self-consciousness is useful, none of us can say.

So it seems to me that there are not only logical but also existential emergents, new things that emerge at points in the phylogeny and ontogeny of living things which it is not permissible, with our present view of science, to regard as predictable. Now it may be that extension of knowledge and enlargement of the categories and scope of various scientific disciplines will result in an increasing unity in the appearance of the world, so that what seems to us now to be absolute emergent will fall into place as an expression of natural law. This would lead to a view of the external world having the greatest conceivable amount of unity. This is a magnificent ideal; one which is close to the heart of scientists – indeed a part of their faith, namely that there is in fact only one science. Such a faith has in the past turned out, as Broad[35] says, to have been much more nearly true than anyone could have possibly suspected at first sight. But with our sciences organized and defined as they are at the present, it cannot be the whole truth about the external world. If one day physical sciences are so modified that they can encompass the whole of nature – an outcome which we may be excused for doubting – then 1 believe it will be found that in the production of this new and as yet unimaginable science, it will be biology which has modified physics and not vice versa.

EVOLUTIONARY PROGRESS

It may today seem a little surprising that the promulgation of the theory of evolution by natural selection in the latter part of the nineteenth century, a period which all are agreed was one of great progress and characterized by belief in progress, should have led to reactions of such violent antipathy amongst so many. I think Bertrand Russell has here placed his finger on an important point when he says that progress, especially during the nineteenth century, was much facilitated by the lack of logic in those who advocated it. This enabled them to get used to one change before having to accept another. [37] Bertrand Russell suggests that when all the logical consequences of an innovation are presented simultaneously, the shock to habits is so great that men tend to reject the whole. If, on the other hand, they are invited to take one step every ten or twenty years they can be coaxed along the path of progress relatively easily. This is a major problem today also and for all of us. Things are happening too fast. The great figures of the nineteenth century, he argues, were not revolutionaries either intellectually or politically. But they were willing to champion a reform when the need for it became strongly evident. It may, he thinks, have been the very cautious temper in its innovators which helped to make the nineteenth century notable for the extreme rapidity of its progress. What was so shocking to the more conservative theologians and the religious people of the time was not only that the principle of natural selection, as Wilberforce argued, was seen to be absolutely incompatible with the word of God. Far worse was the fact that suggesting that there was a relationship between men and monkeys caused the whole system of man's salvation to collapse. This kind of objection was not found only amongst the theologians. Bertrand Russell says acidly that Carlyle preserved the intolerance of the orthodox without their creed – as when he spoke of Darwin as 'an apostle of dirt worship'.

But of course there were many liberal theologians who speedily and gladly accepted the new ideas. It was not too difficult to link the idea of evolution with the idea of progress and to say, with a kind of simple faith, that evolution by natural selection was in fact the creative activity of God seen through the eyes of modern science. Much of this reasoning seems to us now to be shamelessly facile and its neglect of the evils and agonies of the great natural process, viewing them as the relatively unimportant reverse side of that process which was accepted as the law of the world, as rather horrifying. But this attitude of mind in the 1870s was undoubtedly reinforced by the view that not only was progress obvious for all to see, but it would moreover go on for ever. The glorious achievements of the Victorian era were only a prelude. Everything, every day and in every way was getting better and better; and so evolution, to the optimistic, seemed only a plausible generalization from the facts of everyday life. In all this, Hugh Miller's worries over the horrors of the survival of the fittest in the age of reptiles were forgotten. But in due course when men noticed that progress was not quite so inevitable as they had supposed, they came again to look at the view of world history supplied by the fossil record and by comparative anatomy, and were astonished by the number of instances where evolution, far from having resulted in progress, seemed to be retrogressive. Here I think they erred on the side of pessimism and that it is only in recent years that we have been able to take a more balanced view. And it is a balanced view which is here so necessary. It is obvious at a glance that we do not see in the detailed course of evolution as far as it is known, any evidence for an inevitable or *consistent* advance of 'progress'. But though regressive tendencies are apparent to us all (as I have argued elsewhere[45]) it is also true to say that students of the most diverse types and approaches have found in evolution an overall tendency to increases of complexity.[38] We see the development first of increasing independence of, and then control of, the environment. We see increasing elaboration of the animal's central nervous system and associated sense organs. And we find, of course, that these three tendencies very often go together. The picture of evolution in this sense has been vividly portrayed by Teilhard de Chardin in his book *The Phenomenon of Man*. Teilhard was a remarkable combination of palaeontologist of world repute and a mystic visionary, as I consider him, of great power and insight. Since it is not always clear at first reading where the scientist is speaking and where the visionary takes over, some biologists find the work intensely disturbing and irritating. Their reaction to it is, moreover, rendered even stronger by the fact that Teilhard is not in the least interested in natural selection and hardly mentions it – that is, he is not interested in *how* the evolutionary development has taken place, but he is intensely interested in the course which it has taken. And in describing this he has, I think, succeeded better than anyone in making clear the comprehensiveness of the evolutionary viewpoint. He gives a foreshortened picture, after the manner of lapsed-rate cinematography, of the evolutionary process as a whole, placing our own epoch in proper perspective in relation to it. Teilhard's great merit is that he shows us that the real problem is the rise of man; the question of the origin of species is in a sense secondary and, for the most part, far more easily comprehended.[38] By regarding the process of evolution from a sufficiently high vantage point, and with a proper appreciation of the time scale, he is enabled to depict the whole process as one of unparalleled grandeur. Much of his greatness lies in his ability to demonstrate in popular language the existence, in regard to the animal kingdom, of an overall tendency towards increasing complexity and the developmentof mind. He shows in fact that evolutionary progress, if one can take a long enough view, is a reality – an ascending spiral, as he puts it. The essential point in his argument appears to be that we know of no comparable trend in the physical environment, over the geological ages, that could be cited as a cause for this undoubted progress in animal organization. The environments of the Cambrian or Permian could equally well have supported higher types of conscious life, but evidently they did not do so. With the coming of self-conscious life the evolutionary process can be said to have become conscious of itself.

Raven[39] has discussed Teilhard's apparently optimistic views of the evolutionary process, showing how misconceived are many of the criticisms of him. He argues that the modern view of evolution in fact renders denial of progress impossible. Teilhard himself asks, 'Has it even occurred to those who say that the new generation, less ingenuous than their elders, no longer believes in a perfecting of the world, that if they are right all spiritual effort on earth would be virtually brought to a stop?' He continues – 'Surely no Christian can deny that if progress is a myth, our efforts will flag. With that the whole of evolution will come to a halt because we are now evolution.' The sort of despair which is understandably so prevalent just now is, I believe, the result of the phase which mankind has been passing through in recent decades. I myself consider that the scientific world picture of our day does indeed provide astounding evidence for progress. This is not to say that I believe progress to be inevitable in the sense that at any one moment the future is secure. We are indeed surrounded by the most terrible dangers, and at any moment much of what we have gained could be lost and evolutionary advance of mankind, as a social and spiritual organism, put back indefinitely. But I think that if we had been present as privileged observers at almost any critical phase in the evolution of life on this earth – the colonization of the land, the evolution of birds, the first steps in the development of mammals, the early stages of the evolution of human social life – at all times the privileged observer would have been obsessed with horror at the sight of the terrible, indeed almost overwhelming dangers which all these new developments of such promise would have seemed to be encountering. It seems to me that one of the greatest of the many great merits of Teilhard is that he has a more balanced picture than his contemporaries; and that with all his defects, he has been able to display this in terms which the ordinary man can understand. For the first time in the recent history of the intellectual adventures of mankind, I think I would say for the first time in this century, someone has gone far to providing evidence for the correctness of Whitehead's prophecy[40] where he said 'that religion will conquer which can render clear to popular understanding some eternal greatness incarnate in the passage of temporal fact'. No one has come nearer to doing this for modern man than has Teilhard de Chardin. Teilhard wrote.[41] 'Suffering is the consequence and the price of the labour of development.' And *L'ame du Monde* (1918) adds, 'Creation, incarnation, redemption each marking a stage in the divine operation, are they not three phases indissolubly joined in the manifestation of the divine?'[42] To my mind the only conceivable answer to the admitted evil and suffering of the world is in fact the answer Christianity has always given, that creation groans and travails until now. That it is an integral part of the process only acceptable to the mind of man in so far as he begins to see its place in a scheme so stupendous that even the uttermost depths of evil are subject to redemptive action. Teilhard is one of those who has given us in his visionary intensity a transitory glimpse of the future expressed in terms and in a manner which modern man is keyed to understand.

But if we are thinking of mankind as in some sense the crown of the evolutionary process, we must consider his mental make-up at its best and its worst. How far is it plausible and sensible to consider mental development as an emergent from the main process of evolution which has created the sub-human world?

The aspects of human life which it seems to me important to consider in this connection are three: the will, consciousness raised to the level at which full self-consciousness is developed, and the mind of man as a unity. As to the will, I think there is ample evidence of will of a kind among most higher animals. But in the lower animals and indeed in the human baby and only too often in the adult human being, the will is an extremely transient, evanescent and haphazardly directed affair. And I think it can be argued that it is only in the highest aspirations of man in expressing the ineluctable universal obligation to seek and to know higher levels of perfection, to comprehend ever more absolute values, that we find something which quite clearly transcends anything which we have reason to believe exists in animals. And the question arises, since there is so widely established in biological mechanisms a directiveness based on material structure, and if events going on in material structures proceed according to certain laws of causation, how can there be any freedom of the will in any convincing sense of this phrase? Here I will merely say that, if the views resulting from the present developments in information theory which I have outlined are correct, the problem of the freedom of the will, in its physiological relationships at least, takes on an entirely new aspect which will need to be considered afresh both by physiologists and philosophers. It might indeed validate a view, to which I confess I was previously antagonistic, of the type proposed by Ryle[43] which likens the scientific observer to a man watching a game of chess. He knows the rules of the game, and sees that they are not transgressed; but he cannot understand the game because he does not know what is in the minds of the players. As to self-awareness or self-consciousness, we must remember that our whole understanding and knowledge of the external world is deduced from what we consciously perceive. Although we can explain many of the mechanisms which intervene between the body receiving stimulation from an object and the phenomena of consciousness in our minds, still as biologists we have no idea what this awareness means, we cannot see any way in which the phenomenon of awareness could be expressed in terms of anything else. For Waddington,[44] the act of perception and the whole observable world which depends upon it contains an inescapable element of mysteriousness. Considered from the purely biological point of view, the nature of self-awareness completely resists our understanding. It is essentially mysterious. But although I think it is a mystery for biology, it is perhaps not so mysterious for other modes of thought. It is true, of course, that man's methods for apprehending his surroundings must have been evolved from those of animals, and must continuously have been dependent on selection for some kind of efficiency of operation. And for me, as I said earlier, the very fact that consciousness has been evolved is, to my mind, itself strong presumptive evidence that it is not entirely an epiphenomenon. It suggests that at least this aspect of mental life can accomplish something for which the neural mechanism alone, however complex and elaborate, is inadequate.

Thirdly, there is this question of the unity of man's mind. As I have argued elsewhere,[45] there is no doubt that, however much psycho-analytical doctrines may tell us to the contrary, we continue to be convinced that we have in some sense one mind and not many; and that this mind is, in some way at least, ultimately the seat of all types of experience. Nevertheless, in considering the origin of man and the evolutionary story of mind, we find a great deal of evidence to show that there has been, in the course of evolution, a gradual integration of a great many sensory systems, having many different centres of co-ordination; and these, if not amounting to separate minds, seem at least to be independent centres of subconscious activity. Neither in the lower animals nor in fact in the human baby can we find convincing evidence of a persistent consciousness of the object as continuing to exist in the real world. Higher animals do of course achieve this, recognizing an object when they see it again after an interval, and searching for it if it disappears from view behind an obstacle; but it takes the young animal some time to acquire the ability, just as it did when we ourselves were babies. When we consider the lower animals, particularly those in which there is little or no development of the central nervous system (animals such as flat-worms, sea anemones and star fishes) we often get the impression that the animal is only in a very loose and elementary sense an integrated organism. So there is no doubt that such unification as we see in the mind of man is, evolutionarily speaking, a late development and even yet is by no means perfect. But here in particular I think we need a balanced view. The modern developments of psycho-analysis have perhaps led us to place too much emphasis on the divisions and limitations of our minds, and it is indeed clear that the unconscious is as important, perhaps more important in many aspects of human personality, than is consciousness. We can only hope that the human sciences will in due course give us a far clearer concept of personality than we at present have. Neverthe-less to deny that at its highest and best the mind of man is a unity of astonishing security and persistence is to deny some of the most convincing knowledge that we have of ourselves.

CHAPTER 3

Animal and Human Nature

BRAIN AND MIND

Ethology, the comparative study of animal behaviour, provides strong evidence (which it would take too much space to recount here) that something like conscious mind must have been evolved a number of times in the course of the evolutionary history of the animal kingdom. How are we to account for this? Since we can design, even if we cannot construct, machines which should theoretically be able to do all that animal brains can do, and much of what man can do, why is it not more economical to suppose that the appearance of conscious mind in animals is but an illusion; and that the only place in the creation that we are justified in assuming it is where we cannot deny it, namely in ourselves? But I consider that the evidence for some degree of consciousness certainly in the higher animals and perhaps far down the animal scale, is overwhelming. And this (for reasons, which again it would take too much space to describe here) includes, in the higher animals, self-consciousness. So, assuming that there is self-consciousness in the animal world, how did it come about? Is it just some odd epiphenomenon, some strange but necessary outcome of the development of the nervous control system? This I find it impossible to believe. I think we are forced to the conclusion that mind, or conscious awareness, does, in animals, provide some selective advantage over purely unconscious mechanisms or response. That is to say, it would not necessarily have evolved if it had not been of biological advantage at least at some stages in the struggle for survival.[1] Huxley argues that the 'mind-intensifying' organization of animals' brains, based on the information received by the sense organs and operating through the machinery of interconnected neurones, is of advantage for the simple reason that it gives a fuller awareness of both outer and inner situations. It can thus be regarded as providing better guidance for behaviour in the chaos and complexity of the situations with which animal organisms can be confronted, endowing the organism with better operational efficiency and generating quality out of quantity. As illustrating this point, Huxley takes his example from the colour sense. He points out that we know that our different colour sensations depend on quantitative differences in the wavelength of the light received on our retina. We also know that in the optic nerve these different wavelength stimuli are translated into quantitatively different sets of electrical impulses; and yet as all the centuries-old philosophical disquisitions on secondary qualities indicate, we know that colour itself is something qualitatively irreducible. We cannot say that red is more coloured than blue; the sensation of redness is not in any way quantitatively different from that of blueness, it is just different – qualitatively different. And so it is argued that this qualitative difference, though a mystery, is nevertheless a source of biological advantage because it permits readier discrimination between objects. We know from our own experience that it is easy to discriminate between a red and a blue object, but it is much harder to discriminate on the basis of a purely quantitative difference in sensation, e.g. between different shades of grey.

This argument is very alluring. Nevertheless many physiologists would doubtless argue that the essence of the distinction between colours could, at least theoretically, be mediated by a purely physical mechanism – that indeed there *must* be a purely physical mechanism underlying it – and that here again the scientific attitude is to think of the secondary quality as epiphenomenal. But whether or not there is a selective advantage, there is no doubt that, as Huxley says, the animal with the aid of its brain (in the higher animals the organ of the mind) is so to speak 'metabolizing' the raw material of its subjective experience, transforming it into characteristic patterns of awareness which then canalise and direct its behaviour. He calls this 'psychometabolism', and points to the undoubted fact that during the later stages of evolution increasingly efficient psychometabolism is superimposed on the universal physiological metabolism.

It would be dishonest not to refer here to developments in parapsychology for details of which the recent volumes of the *Journal of Parapsychology* must be consulted. The development of the card-guessing technique, due in the first place to Rhine, has led to a mass of evidence for paranormal cognition, some of which seems statistically beyond refutation. This mostly offers evidence for telepathic communication, though evidence for psycho-kinesis (the displacement of material objects by non-physical means) is not to be lightly dismissed [2, 3, 4] and is accepted by some biologists of the highest calibre who have looked critically at the evidence. The reluctance, indeed the refusal, of many scientists even to look the evidence in the face, seems almost pathological – until we remember that it is a commonplace of the history of science that many investigators will refuse to look at facts which cannot be fitted into the existing framework of natural law; or if they look at them, they will prefer some older inadequate 'explanation', however complicated, rather than the simpler one newly presented but violating the generally approved contemporary system.

Two recent trends in parapsychology are worthy of note here. First there is the new and very welcome attempt to relate paranormal performance with recognized physiological states. Secondly, there is the increasing refinement of technique which seems to be providing impressive evidence for clairvoyance as distinct from precognitive telepathy. If this trend is established, and it becomes increasingly difficult to ignore, [5] it will once again open wide the whole question of the relation between mentality, mental perception and physiological mechanisms. If clairvoyance were ever to be indubitably proved it would seem to leave little option save that of the theory of Bergson; according to which there exists some basic cognitive agent or agents, independent of physical mechanism, interpenetrating the physical world to which the minds of exceptional people can at times secure access. If this were so it could well be evidence for considering the brain primarily as a filtering or restrictive mechanism which normally prevents direct contact with other minds, or with an ubiquitous cognitive principle, and in effect normally restricts the minds of men and animals to the data of everyday practical importance presented to them by the sensory systems which have been evolved for the purpose.

Waddington is profoundly impressed by the mystery which is basic to the phenomenon of self-awareness (and indeed at the heart of our whole mental and perceptual life) as completely beyond the powers of biology to comprehend; indeed as an insoluble mystery for scientific thought. I think perhaps here he is unduly pessimistic; although I agree with him in so far as we define scientific thought as including only 'the observation of facts' and the alternating processes of induction and deduction – processes thought hitherto to be the most characteristic of scientific effort. But I think that as we progress, and as we link more and more of the processes of other branches of human activity with those of the scientist, and see the resemblances and congruences between them it is not beyond the bounds of possibility that some further insight into these great questions will one day be achieved. After all there is no ineluctable reason for assuming that the scientific thought of today and of the last hundred years will remain the scientific thought of future ages; nor for supposing that new insights, thought processes and mental techniques may not one day enable us to advance further. Perhaps even new branches of mathematics and types of calculating machines, may in time reveal new relations which are at present hidden from us. No physicist in 1900 could have been expected to foresee the extraordinary advance which the genius of Einstein was about to produce. I agree with Hinshelwood when he says that men of science must not withdraw from the inquiry. The internal external dichotomy must somehow, one day, be removed. But the right grammar and language must first be evolved. In the meantime, studies of human and animal behaviour, sensory and neurophysiology, psychology and cybernetics, communication and information theory, are all to be welcomed as keeping the central problem fully alive until one day *'to the eye of genius some quite ordinary phenomenon will reveal, like the fall of Newton's apple, an undreamt of, but henceforth obvious significance'*.

But to come to earth again: it is true that we must assume some aspects at least of man's mind and of his methods of apprehending his surroundings to have been evolved by selection like his other functions, that is to say, must have been selected for some kind of efficiency of operation. If the kind of approach I have outlined earlier in my references to emergence are valid, we must nevertheless beware of assuming that natural selection alone is governing the whole course and direction of evolution. And I strongly disagree with the assumption implicit in much of Waddington's argument that because man evolved as a physical being he cannot then know any other realms of existence. The animal mind is undoubtedly an efficient instrument for carrying out the essential biological activities; and in so far as man's mind is of the same class of organization, the same thing can be said of the human mind. But I for one should react strongly against any suggestion that we are, let alone must ever remain, in our mental qualities, animals and nothing more.[6]

Whitehead said with profound insight that 'religion is what a man does with his solitariness'. This is indeed the true essence of religion in its highest and purest manifestations. But it is not until we know almost everything about the social life of man that we can begin to understand the nature of his solitariness. To say that man is a social animal gives a glimpse of the obvious; but it is a truism that cannot be too often repeated and no inquiry into human nature can be any use at all unless it is founded on the study of social behaviour. So we need to consider all the evidence we can get as to the origin of man as a social animal; we need to probe into the dark and misty areas where zoology, anthropology and pre-history meet.

THE SIGNIFICANCE OF SOCIAL LIFE

There are two clear and compelling reasons why natural selection should have favoured some kind of social development. These reasons apply far down the vertebrate scale and can be observed in both fish and birds. The fish school or shoal and the closely co-ordinated bird flock, derive advantage from the protection such an aggregation offers when the group is attacked by a smaller number of more powerful and more active enemies. The first manœuvre of a peregrine falcon when attacking a flock of waders, or of a marine predator attacking fish, is for the attacker to adopt tactics which will tend to split up the group or isolate one or two weaker members of it. So the first response both of the fish school and of the bird flock is to mass closer together; not to scatter. The reason for the success of this manœuvre can be illustrated from the problems of defence against air attack in war. Any radar or other type of detector adjusted so as to aim a missile at an oncoming plane is 'confused' if a large number of small objects appear close together. It can only aim into the middle of the mass and not at a single one, and hence is likely to miss. So in nature, a predator cannot concentrate on one until he has got it separated to some extent from the rest. And the ones that get separated are likely to be the less experienced or the weakest. Hence predation of this kind, and the response to it of the preyed-upon species, tends to be eugenic in that there is a strong selection for the good health and high powers of adjustment of those that survive. And if the preyed-upon animal has a means of retaliation and individuals are capable of co-ordinating their response, then again there is safety in numbers. This leads on to the second advantage of social organization in animals. This concerns the finding and utilization of food. If the food sought is not evenly scattered throughout the environment but tends to be aggregated in considerable quantities in a few places, then a species seeking that food is much more likely to do better if seeking is co-operative, since it will increase the chance of the food being found; and when found it will increase the opportunity for its full utilization. Once the find has been made, all can share. If the food animal is one that is able to fight back, has to be subdued or outwitted, once again there is advantage in the social group. Then if evolution is taking the species towards a family organization, where the young are weak and less able to take care of themselves, and where they have to learn much before they are fully competent as foragers or fighters; then again the family or clan will be desirable, in that it will provide a means of protecting the weaker young during the all-important period of their growth and education.

It is now generally assumed (although of course there has been much difference of opinion) that the hominidae arose as offshoots from the primitive Pongidae, or ape stock, some time before the separate evolution of the Cercopithecidae or 'monkeys'.[7] So the study of the monkeys and the apes, the way in which they develop social organization and the nature of the social bond in these animals is right at the root of our inquiry into the social life of man.

The simplest answer to the origin of the social bond is that when feline predators, such as leopards, are about, monkeys are safer in numbers.[8] The hypothesis has been put forward that a three-fold enlargement of the brain in the human stock took place during the Pleistocene. This is regarded as reasonable deduction from present evidence, especially after the recent discovery of *Zinjanthropus* by Dr and Mrs Leakey at Olduvai. This increase in brain size is primarily due to the growth of the cortex and might be put down to one or more of the following developments: (i) the ability to make and use tools, (ii) the necessity to get to know and recognize quickly a considerable number of individuals in the same social group, and (iii) the development of speech. The gap between 'man' and 'ape' has been greatly narrowed in respect of brain size, by the discovery of *Australopithecus* in South Africa and *Pithecanthropus* in Java. The Australopithecines are scarcely above the level of the apes in this respect; and Vallois [9] (1963) considers, from a study of the lower frontal region, that the Australopithecines are more like the apes than men, and that there is nothing to warrant the assumption that they had speech. But they were certainly social and bipedal and were apparently tool-makers. Perhaps this was the only character they possessed among the three characteristics, namely language, tools and war, which are sometimes regarded as definitive of man. There is evidence that [10] *Australopithecus* had a social life little developed from that of apes or monkeys, that they were probably primarily vegetarian, and that the small-brained young could have matured rapidly. But the fact that they were tool-makers and the fact that tool-using and, in a sense, tool-making, have now been observed a number of times in wild chimpanzees by Morris Goodall[11] seems to throw grave doubts on the conclusion of Chance that the development of brain size was primarily due to this characteristic. Admittedly there is a large, indeed enormous, gap between the tool-using of chimpanzees (or indeed the tool-using of the California sea otter and of certain birds) [12] and the tool-making of primitive man. Nevertheless, to attribute the immense increase in brain size for a period of about a million years (fifty thousand generations) primarily to this characteristic seems, to say the least, highly doubtful. If then tool-using and tool-making is not the key to the problem, what light does no. ii on our list throw on the scene? Chance himself [13] suggests that something like an answer can be guessed from the study of present-day macaques and baboons. In these species there is intense competitive behaviour. This is correlated with the breeding premium which implies a very rapid rate of selection of the characteristics that lead to dominance. He points out that the most important of these characteristics that lead to dominance is the ability to control emotive expression at high levels of social excitement. This latter facility, dependent on an enlarged amygdala, would also be a predisposing mechanism for the development of tameness; and so it is suggested that 'tameness' is the basis upon which co-operative social life can emerge in hunting communities. So perhaps it is more plausible to suggest that the clue lies in the development of true social organization. This is organization in which not only are certain individuals dominant, but in which each member of the group must know personally each of the others. Such societies involve, as we find in wolves and many social birds, not merely dominance but submission and leadership, which are much more subtle relationships. On the other hand, the very existence of these faculties in birds and the Canidae gives one pause; because here there is nothing to suggest the necessary connection between a high degree of social organization and a massive development of the cortex or analogous brain structures. It has been argued by the protagonists of the tool-making theory, that bipedal gait released the hand for a more specialized and far more finely adjusted series of activities; and that tool-making is the most likely explanation of the process. Chance and Meade suggest that their control of emotive expressions with its concomitant enlarged amygdala would itself be a predisposing mechanism for the development of tameness; and that this is the basis upon which co-operative social life can emerge in hunting communities. They think it may have been only a phase in man's development but that it probably antedated the period of maximum cortical expansion. It might have served as a necessary preliminary period of selection assisting the background rearrangement of hereditary factors that could predispose man's ancestors for the rapid change that occurred during the Pleistocene. If it were known whether *Australopithecus* had a large amygdala[14] it might be possible to decided whether this type of selection did precede Pleistocene enlargement of the cortex.

As we shall see when we come to consider the development of language a little further, the development of a hunting, as distinct from a vegetarian community, might be a tremendously strong factor in the development of human societies. With the development of hunting, mutual help would certainly become of highly selective value. But here again I think those who have written on this subject are insufficiently alive to the extent to which this already occurs in prehuman stocks. It has only recently been found[15] that wild chimpanzees are sometimes carnivorous and show evidence of precise co-operation in the isolation and catching of *Colobus* monkeys. And after all nowhere in the living apes is mutual co-operation of the degree shown in the hunting behaviour of lions and wolves or the reproductive and defensive behaviour of elephants and dolphins to be found. So as a result of our survey of mammalian behaviour generally, one cannot but feel less and less convinced about the adequacy of any of these three alleged causes of increase in brain size.

I think myself that[16] we may have to look at the matter in a somewhat different light; as indeed did Washburn and co-workers in an earlier and, I cannot help feeling, more balanced assessment of the situation, in which they say 'much of what we think of as human evolved long after the use of tools. It is probably more correct to think of much of our structure as the result of culture than it is to think of men anatomically like ourselves slowly developing culture'.

When we come to man,[16] it is true to say that one cannot discover the cultural capacities of human individuals or populations or races (Dobzhansky) until they have been given something like an equality of opportunity to demonstrate these capacities. If we apply these ideas to the stocks of animals which must have given rise to man, we see that haphazard and probably evolutionarily indifferent characters may have been responsible for quite drastic and perhaps very sudden changes in the earlier stages of culture. Something like cultural traditions can be found even in birds. Modern methods for recording and analysing complex sound patterns have provided definite evidence for what was earlier suspected, that in many species of birds the characteristics of the song pattern vary from one locality to another but remain broadly constant for a number of years for a given locality. It now seems clear that in at least one species, i.e. the European chaffinch, *Fringilla coelebs*, these local variations are indeed dialects, brought about by the fact that young birds learn many of the fine details of their song by singing in competition with their elders during their first year of life.[17] Some of the characteristics of the song pattern for a given locality are thus handed on from generation to generation by imitative learning – in fact a form of tradition. Another remarkable instance is provided by the European greenfinch, *Chloris chloris*. Nowadays this species often eats the seed of *Daphne mezereum*, a small shrub. Although the plant is native to Europe, it is widely grown as a garden ornamental because of its very fragrant flowers. The greenfinches, where the habit occurs, apparently keep the shrubs under observation and as soon as the seed is ready (which is usually about June 14th) descend upon the plant, stripping every fruit from the bushes. Petterson (1961), by means of an extensive co-operative survey, has produced remarkably good evidence that this habit was initiated once and once only between one and two centuries ago in the Pennine district of England, and has spread outwards by cultural diffusion in a fairly orderly manner, northwards and southwards, at an overall rate of between two and four kilometres a year.[18]

Very careful studies have been made of the establishment of new habits, in fact new cultures, in monkeys. This has been done by a group of Japanese workers.[19] The habits of several colonies of the monkey *Macacca fuscata* were studied. In the colony subjected to the most intense investigation, paper-wrapped caramel candies were offered to the monkeys. At first they consistently rejected the candy and only eventually did some juvenile monkeys pick them up and unwrap and eat them. This they did only when not watched by their elders, who in fact attempted to prevent such unusual behaviour. The caramel-eating habit spread first to other juveniles, then to their mothers, then to the dominant males, and last of all to the sub-adult males who are normally kept on the fringes of the colony. This whole sequence took at least three years, and even then not all the sub-adult males had developed the new habit. This[20] provides a striking analogy with the food habits in different human populations. What one eats or refuses to eat is determined largely by what other members of one's family and one's group regard as fit or unfit to eat. This of course does not rule out the occurrence of genetically determined or conditioned idiosyncrasies in dietary matters – it merely shows that very great changes can take place and persist, at any rate for a short time, even though there is no genetic base for them. The Japanese workers also studied and compared several bands of *Macacca fuscata* and found them noticeably different not only in food habits but also in social structure and inter-individual behaviour. In one of the colonies a new food habit (eating wheat) was introduced by the dominant male leader of the band and spread rapidly to females and to juveniles. In another band, the precedence of dominant individuals was enforced by fierce fighting and most individuals bore traces of wounds; while in still another the fighting was merely 'symbolic'. In some colonies the foraging for food started from a fixed centre of operations, just as it does in greenfinches, but other bands wander around and rarely return to the same place. Some bands of monkeys have freer sexual life than others; a female who copulates with a sub-dominant male may be either severely punished by the dominant one or tolerated by him. The only reasonable conclusion from this work at present seems to be that different monkey colonies have different learned proto-cultures.

With this kind of example in mind, let us consider again the evidence for cultures in early man. Schultz has an ingenious suggestion[21] as to the way in which the attitudes of subhuman primates might have given rise to some of the most characteristic examples of early human cultural traditions. He points out that every zoo and primate laboratory has witnessed instances of the pitiful devotion of simian mothers to their sick infants, and knows the long struggle to take a dead or even decomposed baby away from its mother, who will clutch the corpse to her breast day after day. Sick or wounded juvenile or adult monkeys which are incapable of keeping up with the group, search for a hiding-place perhaps in a cluster of dense foliage or, if they have fallen to the ground, in the deepest niche amongst roots or rocks. If they reach a cave they will soon be found in the farthest darkest corner. There they remain, still and quiet, until they recover or die. It seems that, like monkeys and apes, the early hominids withdrew to the best available hiding-places as soon as they became too weak, from one cause or another, to live with their groups. It was not until the later hominids became encouraged by the light and protection of fire, that they began to use caves, instead of trees, as a nocturnal retreat for those healthy individuals who could be sheltered thus. It is plausible, he says, that the primary role of caves for primates connected with illness and death may very well have influenced the later behaviour of cavemen; especially their attitude towards skeletons found in their innermost recesses. Acts of ritual anthropothagy[22] seem to have been associated in a number of cases with a cult in which the occipital hole of the skull was enlarged with blows from a club. The brain was then devoured, presumably by those who wished to assume the virtues and merits of the dead man.[23, 24] Similar anthropophagous tribes in central Africa apparently practised mutilation rituals for a long time. Thus the presence of broken jaws and skulls indicate a funerary cult. What can be the significance of this? What did these men think about before a dead body? What could be the meaning of the rigidity, the cold and impressive stillness? The dead man is there, but he does not speak, he appears not to see or hear, he is all the more formidable. He appears to be jealously guarding a terrible secret which none can take from him without assuming part of his substance. So it is natural to think that the organ of command, so expressive in life and now fixed in rigidity, should be an object of special veneration; and so they carried a few bones of the deceased about the person like a talisman. And so it seems as if the kind of behaviour that occurred in *Zinjan-thropus* was retained, or again developed, without substantial modification through hundreds of thousands of years. It is all highly conjectural; but it is hard not to believe that during these millennia secret psychical activity was ceaselessly exercised. That cultures of such persistence must have very powerfully affected the pressures acting upon the individual cannot be doubted. Perhaps we shall never know enough about the cultures of prehistoric man and his forebears to be able to get the picture very much clearer than it is at present; but it is plain that ritual mutilation of the base of the skull was performed by early and late Neanderthal man for a period estimated at about a quarter of a million years.[23] The stability of this and other evidences of culture can only be explained in terms of the power of tradition, and of cultural continuity through evolving human races which, meanwhile, changed body type and most certainly changed genetic constitution. I think that the ethologists, taking a wide view of animal behaviour, would tend (perhaps surprisingly to the anthropologist) to emphasize, in cultural development, the psychic factors as more important than any other one feature in determining the evolutionary course of the hominid stock. But there is one point which I think deserves emphasis at the conclusion of this section. It has often been pointed out that modern man as an individual is weaker than at least many of his ancestors, and that biologically one of the most striking features of mankind is the helplessness at birth and for the long period of childhood. Why did selection lead to the formation of a species made up of individuals of a more infantile and therefore less robust structure? A Russian anthropologist, Roginski, answers this question by saying that it is just this weakness of the individual that favoured the establishment of a more organized and more substantial collective group in which each of its members is first of all seeking protection at the core of this collective group, and in the second place is capable of submitting to its exigencies.

Once we have reached the stage at which family groups are led by some means or another to the development of a clan structure, then as we see from a study not merely of monkeys and apes, but also of birds and many social mammals, very human characteristics begin to appear. As soon as we have individual affection and longlasting attachments, as in wolves and in birds such as geese, we begin to see traces of a jealousy which is quite astonishingly similar to that vice in humans. In this there is no strictly scientific *experimental* work on which to base one's conclusion: such evidence as there is tends therefore to be anecdotal. I think, however, it may be justly said that the more one becomes acquainted with the social behaviour of higher animals, the harder it is to explain away what appear to be evidences of jealousy. I believe it would be hard to find an experienced dog owner who had any doubt that something like jealousy occurs in the dog's mind; and there is a considerable amount of evidence for wolves.[25] In the case of primates there is a remarkable account [26] of a rhesus monkey which, when under a state of intense stress due to sexual deprivation, took to quite savage self-mutilation, biting his hind feet and tearing huge jagged rents in the muscles of his legs. The account gives the impression of feelings of intense jealousy experienced by the monkey; but one must be on one's guard against accepting some parts of this account at its face value in view of later criticism of the case.[27] The most remarkable instance of what is apparently jealous behaviour amongst mammals that has come my way was observed in my own laboratory by my colleagues Professor R. A. Hinde and Miss Y. Spencer-Booth. We have a number of groups of rhesus monkeys each consisting of a male, a number of females and the resulting young, each group being housed separately in a compartment of a monkey house consisting of a large outdoor enclosure approximately 2,000 cu. ft. and an indoor shelter communicating with it by a swing door which the monkeys can work themselves, and a glass window. When a baby is born in one of these groups it is always the object of intense interest on the part of the various 'aunts' and their possessiveness may give rise to a good deal of squabbling. In the latter half of 1962 a female Rosie, at the bottom of her group's hierarchy of females, produced a baby. This infant seemed particularly attractive to Eliane, the 'favourite' wife of the male Tom, and thus at the top of the hierarchy. Eliane actually succeeded, on two occasions, in getting the baby away from Rosie and much effort was needed to restore it to its rightful owner. When Rosie had finally re-established possession of the coveted infant, Eliane was seen on many occasions going through a remarkable performance. When Rosie and her child were 'indoors' looking out through the window, Eliane would come and gaze at them from outside. Sometimes she would 'shadow box' with Rosie through the window and would then run away squealing as if she had been attacked. This squeal is normally only used when a monkey is in danger of attack. The result was, as might have been expected, that Tom, the male in charge, went in and beat up Rosie. It is difficult to avoid the conclusion that this was a carefully thought out piece of behaviour 'designed' by Eliane to ensure that Tom punished her rival.

Amongst birds the most remarkable case that has come to my notice concerns a pet barn-owl (*Tyto alba*) which belonged to Dr Miriam Rothschild. The barn-owl was hand-reared from the nestling stage. At the time of its capture the owner was ill and the bird lived in a waste-paper basket at the end of the bed. It was therefore constantly in the company of its foster-mother to whom it became extremely attached. It was never pinioned and lived free in the house. It was also house-trained and returned always to the same perch on every occasion. Dr Rothschild describes the bird's nature at this time as catlike – very independent but affectionate and sensual. It particularly liked being stroked and tickled, and it would sidle up to its owner (say on a writing-desk) and solicit stroking and fondling. It would then close its eyes and show every symptom of enjoyment, swaying and 'purring'. When the owl was seven months old, in 1945, the owner went to London for six weeks while her child was born. She returned with the baby in early February. The owl recognized her and showed the same affectionate disposition as when she first saw her. But directly the bird saw the baby, a sudden change came over its attitude. It savagely attacked the owner and from that moment onwards was unsafe. Dr. Rothschild found it difficult to believe the situation, for she says this greatly beloved bird was the most affectionate tame animal she had ever possessed. Every effort was made to win back the owl. She continued to keep it in the living-room and wore protection on her head and face. The owl never had the chance to get near the child and only rarely saw it, but its hostile attitude did not change. It was not nervous or afraid, but continued to launch unpredictable and savage attacks on the owner, *but on no one else.* The bird was, for instance, docile and affectionate to the cook. If the owner approached it with food it was quite possible that the owl would strike savagely at her with its claws and beat at her with its wings. After trying for six weeks to win back the bird's affection, it was decided with extreme reluctance to part with it. On separation from the foster-mother this owl made a "nest" and laid sterile eggs. The story ended a few weeks later after some fertile eggs had been provided for the owl to sit upon, when some well-meaning but ignorant visitor gave the bird a moribund mouse to eat. It turned out that the mouse had in fact been poisoned and the owl died too.

FREUD AND THE EVOLUTION OF MIND: A BIOLOGIST'S CRITIQUE

As soon as the complexity of organization of animal societies was realized by psychologists, the way was open for speculation and comparison aimed at linking the ideas of the psycho-analyst, primarily Freud, with those of the biologist. This is a field in which facts are so meagre that speculation easily runs riot (and has often done so); although one should say at once that there are many very sound and useful comparisons made between the work of the psycho-analyst and the ethologist.[28] Nevertheless it seems advisable to be fairly outspoken about some of the theorizings of Freud as looked at from the ethological viewpoint. Freud himself was trained as a zoologist and might have been expected to be somewhat more critical in his approach along these lines. Perhaps the explanation is that his biology was very much that of his training and early work as histologist and anatomist rather than as a student of living animals.

However this may be, one cannot get very far in the process of considering the transition from animal to human behaviour without encountering the theories of Freud; and in my view his greatest contribution to the understanding of human nature resides in two major conclusions. The first of these comes from the overwhelming evidence he provides for the fact of subconscious and unconscious mental processes of great complexity. The second comes from the way in which he was able to relate the innumerable instincts and drives with which previous writers had saddled humanity, as linked in a very fundamental way with the sexual instinct alone. It is with this second achievement that I am now concerned. It is true that Freud often speaks, for example, about the ego instincts as opposed to the sexual instincts and about the life instincts as comprising instincts of self-preservation, etc. He does not, as far as I am aware,[29] discuss any of these other instincts in detail; instead he fastens securely on the sexual instinct as in some way the basis or condition of all the others. For Freud, instincts are innate appetites or urges arising internally, and so involving a specific energy. These instincts are qualitatively distinct in feeling, and the 'energy' which drives them, aims at satisfaction by means of objects in the external world[30] and comprises all these instinctual levels primarily at the level of the unconscious. This striving for satisfaction aims on the one hand at reducing tension and excitation, and on the other at achieving satisfaction. It is clear to anyone with any experience of the structure of a monkey clan, however superficial, that the sex instinct is certainly the predominant driving force of the life of the community; and it is central to the formulation of Freud that a large part of the basic conflicts and difficulties of human life are associated with the long period of human childhood during which the instincts are

79

present in the conscious and subconscious but are denied their expression in overt behaviour. This obviously has some close similarity to the approach of the ethologists. In ethological language we should say that appetitive behaviour is present but that circumstances prevent the achievement of the consummatory act; and that this inability to achieve the 'goal' gives rise to what the psychologists call tension, a process which is revealed in animals by the tendency to perform displacement activities.

In Freud's system the core of our being is formed by the massive unconscious id. This is regarded as the structure which gives rise to the internal drives. The id is cut off from the external world; it remains centred in 'its own world of perception'. It obeys inexorably the pleasure principle, and itself knows no inhibitions or anxieties. The ego, according to Freud, is a differentiation of the id brought about by the activities of perceptual consciousness. It is that part of it which has been modified by the direct influence of the external world; and it has the task of bringing the influence of the external world to bear upon the id and its tendencies, endeavouring to modify the pure pleasure principle of the id in conformity with the reality of the external environment. Thus, to quote Freud, 'the ego represents what we call "reason" and "sanity" in contrast to the id which contains the passions . . .' The ego learns how to manipulate the external world in order to satisfy the appetites emanating from the id. It also perceives and then tries to control the instincts. It proceeds, that is to say, from obeying instincts to curbing them, in accordance with the demands of the external world. Only part of the ego is conscious; the rest of it extends into the unconscious and into the massive id of which (so it is paradoxically supposed) it is its origin and differentiation.

The super-ego is held to be a later differentiation of the ego itself, and to be the outcome of its need to accommodate itself to its early social environment and family situation[31]. In some respects the super-ego comes to constitute the conscience. Once it has been established, its demands upon the ego no longer come from outside but from within the individual's own personality; and so the super-ego watches over the activities of the ego, inhibiting certain of its actions which are undertaken in the service of the id. Rather paradoxically, the super-ego is thought to extend farther into the unconscious than does the ego, and it is therefore to a large extent inaccessible to the latter. Because of this close contact, there is a free communication between the id and the super-ego. Up to this point at least Freud's system is clear and plausible, although I think the acid comment of Zangwill is justified, that the super-ego is that part of the ego which is soluble in alcohol! But as Freud goes on, he proceeds to argue that the super-ego is to some extent phylogenetically determined. And it is here that Freud's strange ignorance of and insensitivity to the biological approach becomes manifest. Thus in fact Freud[31] argues 'that man is descended from a mammal which reached sexual maturity at the age of five, but that some great external influence was brought to bear upon the species and interrupted the straight line of the development of sexuality. This may also have been related to some other transformation of the sexual life of man as compared with that of animals, such as a suppression of the periodicity of the libido and the exploitation of the part played by menstruation in the relation between the sexes.' Much of this is again reasonable; but the biologist has however to part company with Freud when, in developing his ideas about the nature of instinctual conflict he found it necessary to introduce two conflicting instincts, the life instinct and the death instinct. Freud, in his *Beyond the Pleasure Principle*, developed this very tentatively, merely as an interesting line of thought, and at the end of the book he explains that he is not convinced by the conclusion he has come to, though he thinks the arguments would be worthy of careful consideration. Nevertheless,[32] in his later writings he speaks as though he does in fact seriously accept his conclusion and the argument, and he utilizes it in various connections. He goes even further and implies that the death instinct is present in all animal species throughout the entire organic world, and to do this is simply to write biological nonsense. What the psycho-analyst has to say of importance on the subject of instinct is solely with reference to the part played by instinct in *human* life. Those generalizations which he makes with regard to instinct in animals or in organic life in general are practically worthless. Indeed there is a great gap between instincts as discussed by psycho-analysts and instincts as discussed by the other theorists. 'We can now see that – when the life and death instincts are discarded – there is no such gap at all.'[32]

The same kind of criticism can be levelled against some other elements in the classical psycho-analytic system of explanations. Another of these is the birth trauma. Infants delivered by caesarian section[33] do not seem to develop very differently from normally born ones; yet caesarian babies should be free of such traumas, or else suffer traumas of a different kind. Gesell and Amatruda[34] compared the behavioural development of babies born about a month prematurely with that of normal ones, they found that all reached similar developmental stages at the same age, counting from conception rather than from birth. Since premature babies have the 'advantage' of being exposed longer to socializing influences, there should be clear difference in the course of development; but it is evident that in fact such socializing influences have to come at a certain stage of the bodily development to be fully effective. (This conclusion as to human beings ties up very nicely with the observations of ethologists upon the process of imprinting.) Finally, it seems unlikely that a process biologically as indispensable as childbirth would result in the traumatization of the whole species? Granted that the painfulness of childbirth is one of the biological disharmonies of human nature – though mainly in civilized cultures. But, in any case, should not a species saddled in addition with birth traumas have become extinct long ago?[35] Similar criticism can be levelled against the idea of the Oedipus complex. As Dobzhansky points out, to Freud, who was a believer in the inheritance of acquired characters, it may have been a credible assumption. But in addition to this Freud believed that the evolutionary history of the species is repeated in an abbreviated form in the development of every individual. He consequently assumed that all human children pass through the Oedipal phase.

I think there is much justice in the criticism of Williams[36] which, while admitting that Freud's genius led to the discovery of a completely new system of explanation, argues that it resulted in a system that cannot be apprehended merely by intellectual study. He likens him to St Augustine and says, 'as with Augustine, so with Freud ... there is a demand that the risk be taken of opening oneself to a greater reality than is at present known to us. For Augustine the reality was God: for Freud, the unknown self. For Augustine, the way was prayer: for Freud, analysis.'

SOME CHARACTERISTICS OF SOCIAL LIFE

But although for the biologist the system of Freud has severe, indeed fatal, limitations; once we get outside the basic ideas of conflicting unitary drives as an explanation of animal behaviour, and cease attempting to explain the origin of human nature by such biologically fantastic hypotheses, we can find many significant and useful similarities between the social behaviour of animals and men, particularly in relation to subjects such as war and aggression. This has been done very convincingly and dramatically by Konrad Lorenz.[37]

I have already referred to the biological function of a closely co-ordinated swarm or troop – namely that of being able more effectively to counter the determined attack of a large and fast predator on a single individual. According to Lorenz, and I think all would agree with him, the anonymous swarm or troop as seen in schools or shoals of fish, and probably also in certain very closely co-ordinated flocks of birds, is the primitive form of social life. In the true swarm or troop of this kind there are no leaders and no followers, in fact no individuality. The whole group is co-ordinated by fixed responses to releasers of one kind or another – releasers which have been evolved over long periods by the action of natural selection. Such a troop is the primitive foundation of social life. The more numerous the individuals, the more intense the herd cohesion and herd 'instinct'. There is a very illuminating experiment by Professor von Holst in which operations were carried out on the brains of minnows. The result of this was that the normal social responses were eliminated. Consequently those fish with no social responses became the leaders, since they were responded to but no longer responded themselves. Previously, in the natural school, as in the ant swarm, there were no leaders. In the primitive social organization of this kind – that is, in troops or schools or swarms – any one comrade is as good as another, and there is not the slightest evidence for supposing that there is any individual recognition amongst them. Up to a point, it can be the same in human society; a good business partner is essentially one with the same business interests – not necessarily one with acceptable or agreeable ideas on general subjects.

The next point to remember is that all animals with bond behaviour also have aggressive behaviour. The result of this is, in these animals with more primitive social organization that we have been discussing, the phenomenon of 'individual distance'. Where there is a constant stress between the tendencies to scatter and to keep contact, the former mediated by the innate releasers and the latter by the aggressive tendency, there is a point at which the two normally balance, and that is the 'individual distance'. It can be seen not only in fish but in many bird flocks, in the swallows perched on the telegraph wire or the seagulls on the harbour wall. Thus in the fish shoal or the anonymous flock of birds, all is relatively simple and controlled. In birds during the breeding season, however, any one individual has to maintain much closer contact with one other individual, namely the mate. That is to say, in many species the breeding contact is closer than the contact in shoal, swarm or flock, and so something has to be done to assuage the aggression which otherwise would manifest itself and break up the breeding cycle and organization. Lorenz shows that the fury released by the female (first viewed as an intruder) is exploited by redirection to attack not the female but rival males. This has been called 'abreaction', and is thus primarily a means of getting rid of aggression. As between the male and female, the aggressive reaction is not only dispersed by abreaction; it is also checked or limited or dispersed by greeting ceremonies and appeasement ceremonies at the nest. Nevertheless, it is still there, waiting to be redirected against the rival male. When, however, one comes to mammals and to those birds in which individual recognition is fully developed – that is to say in which the individuals recognize one another personally, and in which long-maintained attachments develop, then the social organization is different. This happens in the geese, and in many mammals where the family clan or group, most or all of the members of which are known to one another personally, becomes the basis of the organizations. Mammals, in contrast to birds, are contact animals with few devices for the maintenance of individual distance. In fact they tend to huddle. Therefore if strangers are introduced into the situation, as in a cage, in which they cannot flee from the owners or cannot hide, disaster will supervene. The same is true to an even greater extent of domesticated or tame animals, introduced into a wild stock; since these domesticated individuals may have lost (or may never have developed) the appropriate appeasement responses – which we see in dogs in the form of cringing or presenting the neck, that is to say the most vulnerable part, to the dominant animal. Thus such animals artificially introduced may be in worse case than wild strangers introduced into a clan. Lorenz describes the terrible lot of a rat, artificially provided in a cage with 'foreign kindred'; it will be slowly torn to pieces by its associates.

It is quite easy to imagine that in the kind of society which many of the ancestors of man may have passed through, hate may have been a good thing. Hate may today be a good thing in an animal society as long as it does not result in fights amongst kindred.[38] The hate of an animal society for strange clans or groups of its own species may in fact be ethologically eugenic from the point of view of the selection of characters which go to develop a highly elaborate and secure social life. Thus it may be that the greatest social, moral and religious problem of today, namely that of war, cannot be fully understood without recourse to a knowledge of the progress of the evolution of human social life.

I referred above to the fact that essentially the same problem which we have just been considering in mammals also arises in many birds during the period of close contact while breeding; and in some birds such as the geese, which have societies more on the lines of the mammalian type, during most if not the whole of the life cycle. In geese, as every observer of a farmyard flock will have seen, the gander, having dispersed its aggression by an attack on a rival, returns to the female and by its triumphal display further cements the bond between male and female, rendering to that degree easier, on future occasions, the further dissipation of aggression by abreaction. There seems no doubt that many of the complex appeasement displays which we find in birds have been evolved in fact from redirected aggression. With a little licence one can just say that a properly civilized goose sublimates and socializes its aggression, consolidating its family pride and tightening its family bonds thereby. One of the most usual experimental techniques of the ethologist is to rear animals in complete isolation from their kind. Such isolated animals are known as Kaspar Hausers, after a German legend of an isolated human child. It is characteristic of these Kaspar Hauser animals, whether birds or mammals, that the state of aggressive tension is heightened as a result of this isolation in a manner such that when the male animal (or indeed a female, for that matter) is provided with a mate after this long period of isolation, although the mating response may be heightened, the aggressive drive is out of control. Indeed this excessive aggression can be the greatest source of difficulty in breeding hand-reared animals in zoos. They cannot disperse or redirect their aggression. In the hen domestic turkey, with eggs or young, the sight of a furry object is the innate releaser for violent attack. This is an effective means of ensuring that mice, rats, or weasels which might do harm are unrelentingly attacked. Now young turkeys also look furry or fluffy. Attack on them is prevented by the peeping noise which they make; this noise has a powerful inhibitory effect and prevents attack. But if we make a turkey deaf, it kills its own chicks immediately they are dry because they are now fluffy and release the aggressive drive and the hen – being deaf – cannot hear the protective peeping.

Lorenz has some particularly relevant observations on a situation which can arise quite easily when one is keeping geese in numbers – that is, the formation of homosexual gander pairs. Such pairs can survive indefinitely if there are, as in normal pairs, sufficient opportunities for abreaction to lead off the aggressive drive. But if a display between two homosexual ganders once leads to a real battle, the two can never afterwards make it up; since they know each other personally, and a *personal* hate supervenes. In the flock they keep away from one another and look the other way and do everything in their power to avoid an encounter. Perhaps man's chief difficulty today is to be found in the fact that he cannot manage any of his fundamental instincts because he has modified his environment so fast that none of his instincts fit in with his new circumstances. This is another aspect of the fact, mentioned in the last chapter, that culture changes faster than gene pool. It is obvious that conclusions of this kind may be immensely significant for man in some of the circumstances with which he is now contending. I shall return to this subject in a later chapter.

The characteristics of play in animals have already been referred to above. Play is a complex phenomenon and hard to define. In the higher animals it is extraordinarily widespread; but we should not expect to find much evidence of play in invertebrates because not only is their behaviour simpler and more rigid, but also because – as all the evidence goes to show – the link between appetitive behaviour and consummatory act is much stronger there than it is in the higher animals. There is, so to speak, none of the looseness of organization in the lower animals which enables their activities to emerge in play. True play may be expected wherever the circumstances of environment or life history are such that appetitive behaviour can become emancipated from the restriction imposed by the necessity of attaining a specific need or goal; in other words, in the young of animals with a prolonged post-embryonic development and where the primary needs are satisfied by the care of the parents (e.g. in nidicolous birds and in mammals – so that the young are not yet involved in the serious business of survival). In the normally activated appetitive behaviour a new obstacle or hindrance will of course draw the attention of the animal to that obstacle and as a result it will learn something about it and about ways of circumventing or avoiding it. This is the process by which the world of an instinctively motivated animal becomes delimited, and it is the way in which behaviour often becomes directed and perfected through learning. But the instinctive drive is not thereby lessened – on the contrary, its threshold may be lowered. If then the goal of the instinctive drive is satisfied either by the abnormally easy conditions of life as in domestication, or by the attentions of the parents or the care of a human master or keeper, then (provided the appetitive behaviour is not too strictly tied to the consummatory act but has so to speak appropriated some motivation of its own) we get the beginnings of a general exploration of the environment which may often take the form of play. This is particularly likely to happen in species where the bodily organization is such as to allow the manipulation of discrete objects or things in the environment. The process of play can thus lead to a relatively enormous widening of the animal's horizons. Provided then that the conditions of life are easy the practical value of play may lie in the means it provides for learning about and mastering the external world. But it may tend to become more and more the main outlet for the animal's energies. If this happens, an element of danger is introduced, since hindrances and obstacles are no longer simply objects to be avoided or surmounted but may become objects of investigation in themselves, and behaviour may become too much separated from the concerns of daily life. But just because the instinctive limitations are now in a measure transcended, new worlds are opened up and new freedom achieved. Thus possibilities for large and novel evolutionary developments are provided.

When we come to the higher birds and mammals with a family or social organization, a particularly important aspect of play becomes evident. In this play all the social inhibitions are maintained even when the play is violent and passionate. It may on occasion change into serious fighting, particularly when a predatory animal such as a cat or ichneumon is playing with real prey. But normally, as in the playful combat between two young dogs, there is a clear and obvious difference from serious fighting. However fierce they may appear, they do not bite each other seriously. We can only say that they are not in earnest.

The studies of Bally[39] in Switzerland show how closely similar in many respects is the play of man and the higher animals. There is indeed good reason for thinking that the prolonged childhood of the human species, coupled with the extreme infantile sexuality followed by a latent period, occurring as it does so long before there is any possibility of consummatory sexual behaviour, has been of prime importance in the process of freeing appetitive behaviour from the primary needs. It seems hard to escape the conclusion that this has been the path by which man has come to be an artistic and aesthetic animal; though as we shall see there is now plausible reason for thinking that something like true art, at least in its first glimmerings, is shown by birds as well as mammals.

Birds in fact show artistry of a type which leads both to visual and auditory satisfaction. The Bower-birds of Australia and New Guinea build display grounds consisting of variously constructed bowers to which they eventually entice females and in or near which mating takes place. Nests however, are built away from the bowers. These bowers are decorated in various and often highly elaborate fashions; some with an avenue approaching the bower containing objects such as bleached bones, pieces of stone or metal or (if they can get them) shining coins or other objects produced by man. Some will arrange brightly coloured fruits or flowers which are not eaten but are left for display and replaced when they wither. Yet other species paint the walls of the bower with fruit pulp, with charcoal or with dried grass; and at least one manufactures a painting tool out of a small wad of spongy bark. Some species not only select and maintain in an attractive state objects that seem beautiful to us, they stick to a particular colour scheme. Thus a bird using blue flowers will throw away a yellow flower inserted by the experimenter, while a bird using yellow flowers will not tolerate a blue one. It seems impossible to deny that a well-adorned bower may give the bird a pleasure which can only be called aesthetic. This investigation of what may be called proto-aesthetic phenomena in mammals and birds has been followed up by Rensch[40] who offered experimental animals (monkeys and birds) choice of pieces of white cardboard, differently patterned, as materials for play. The animals showed preferences for symmetrical and rhythmical patterns, and for those with steady rather than faltering lines. We get the same impression from much precise modern work on the analysis of bird song. As Szöke has argued, birds with the most highly developed song show a general similarity in the structure of their tonal system. This seems to have been originally determined by the peculiar nature of the vocal apparatus of the primitive birds, and the overtone series which it can produce. Within these inherent limitations some of those few species of birds which have as yet been critically investigated show evidence of spontaneous rearrangement of phrases and the invention of new material which suggests something similar to real musical invention. [41, 42] How right Robert Bridges was when he said:[43]

'Verily it well may be that sense of beauty came
to those primitiv bipeds earlier than to man.'

Indeed, it is now seriously argued by Szöke[44] that in so far as
man uses intervals of the natural series of overtones, these must
have been learned from the environment – since they are not
necessitated by the nature of our vocal equipment. It is perhaps
plausible that the intervals which are acceptable to the human ear,
as normal and natural for music, are in fact those intervals which
were first offered to the ancestors of man by bird song. Other
animals do not have much in the way of song; but the fundamental intervals of human and bird song are the same; and
highly-developed bird song was audible at man's first appearance
in time. Since man always had bird song all around, impinging on
his ears, is it not reasonable to suppose that he developed a musical
signal system by imitating the birds? This bird song was palaeo-
melody. Human pre-music gradually developed, according to this
theory, to be the component of the total art of primitive society
and later to be a differentiated musical art; always – to quote
Szöke – 'in the closest relationship to the development of science,
under its influence and in its interest'.

By comparison with these achievements of wild birds, the
artistic abilities of subhuman mammals seem meagre in the
extreme. I am indeed impressed by the sense of unity and of design
shown by some of the 'paintings' which chimpanzees can be induced to produce in captivity. However, they are to our eyes
feeble in comparison with some of the apparently biologically
redundant performances of the bird vocal organs as they appeal to
our ears. I shall be more impressed by the evidence for ape art if I
learn of examples of its occurrence in the wild. It may be that the
drawings of chimpanzees have been, in some subtle and at present
undetermined way, affected by the unconscious predelections of
the experimenter (in the same way as the famous 'calculating
horses' were in fact responding to the unwitting signals given by
the trainer). Perhaps, too, the order in which the painting

materials were presented to the apes, and the way in which this was done, may have affected the result. I am impressed by the artistic efforts of apes; but I am not convinced that they offer such clear examples of the birth of true artistic ability as do some examples of bird behaviour and especially some recordings of bird song or, coming again to mammals, as do some of the experiments of Rensch.

It is again plausible to think that some at least of man's artistic activities, notably his dances and displays involving head-dresses and other ornaments, may have been stimulated by watching the ceremonial displays of birds. Here perhaps the really fundamental problem is why, in so many cases, the patterns which have been produced in evolution as recognition marks for sexual behaviour (one has only to think of the magnificent colours and patterns of many butterflies and the display plumes of many birds) strike us as beautiful? Does it imply some fundamental unity between the mind and perceptual systems of groups as far apart as the Aves and mankind?

CHAPTER 4

Animal Societies and the Development of Ethics

ANIMAL COMMUNICATION AND LANGUAGE

Directly we attempt an imaginative reconstruction of the kinds of community which, in the different stages of their development, the ancestors of man must have passed through, we find ourselves wondering about the means of communication employed, and speculating as to the origin of language. It was fashionable at one time to assume, as a matter of course, that amongst all the differences between man and animals the possession of language was the most significant. But whatever the truth of the matter, it is clear that we cannot today accept this conclusion as self-evident; the validity of a division of this kind requires the closest scrutiny. Let us then look at the communication systems of other animals and try to evaluate both their similarities and their differences.

The methods of communication employed by animals are very various. Visual signals and displays or gestures, from the dances of bees to the displays of birds, odour trails and the marking of territory by excreta, and the secretion of special glands – all these are common and widespread in the mammals and the social insects.

In many mammals secretions largely take the place of visual signals for inducing changes of mood. However, when we come to the higher mammals, which have something like a clan organization (such as the wolves and still more the primates) we find visual signals of great subtlety. Wolves have at their

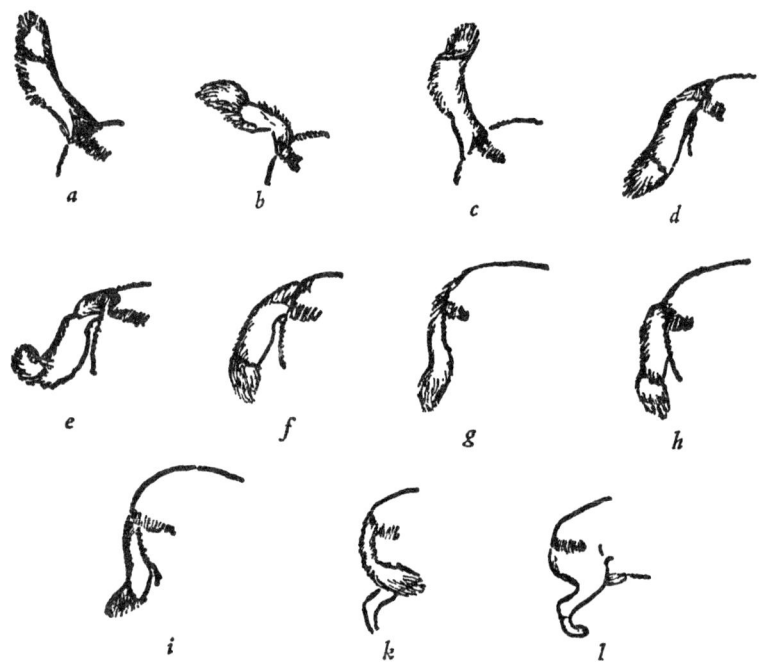

FIGURE 4 Use of the tail for expression in the wolf

A Self-confidence in social group B Confident threat C With wagging; imposing carriage D Normal carriage (in a situation without special tension) E Somewhat uncertain threat F Similar to D. but specially common in feeding and guarding G Depressed mood H Intermediate between threat and defiance I Active submission (with wagging) K and L Complete submission.

(after Schenkel, 1948)

command at least twenty-one communicatory signals, of which fifteen probably involve some visual elements; the others being olfactory and tactile. They may take many forms, but the face is particularly expressive; and in these signals, while the movements and attitudes which bring them about are probably largely innate, recognition of their meaning has no doubt often to be learnt by

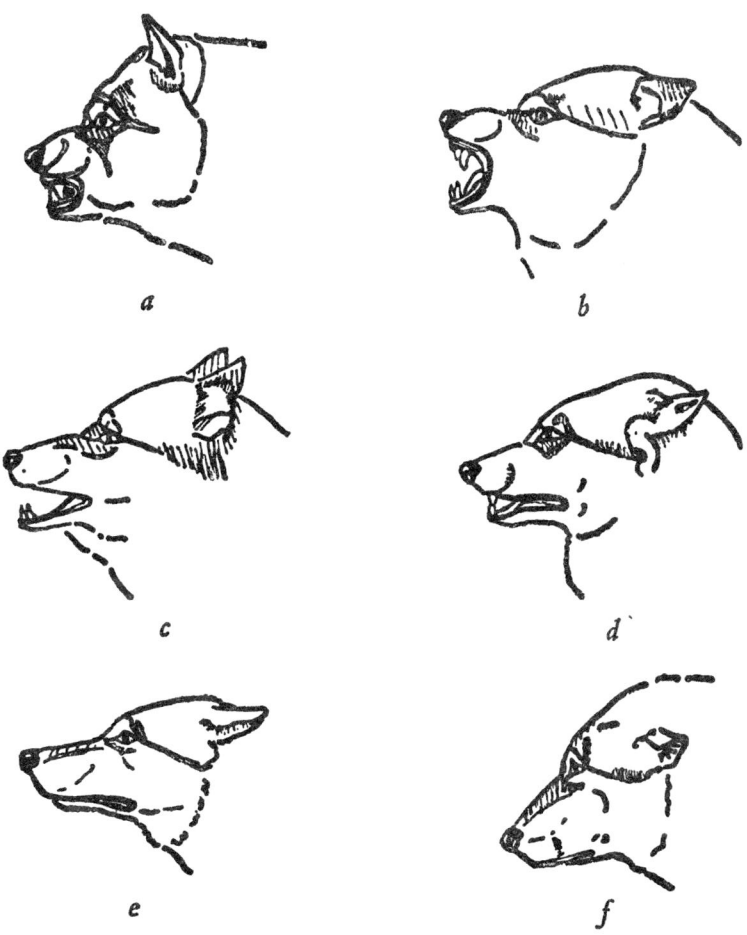

FIGURE 5 Studies on the expression of emotion in the wolf

A Fully confident threat B High intensity threat with slight uncertainty
C Low intensity threat with uncertainty D Weak threat with much uncertainty
E Anxiety F Uncertainty with suspicion in the face of an enemy.

(after Schenkel, 1948)

95

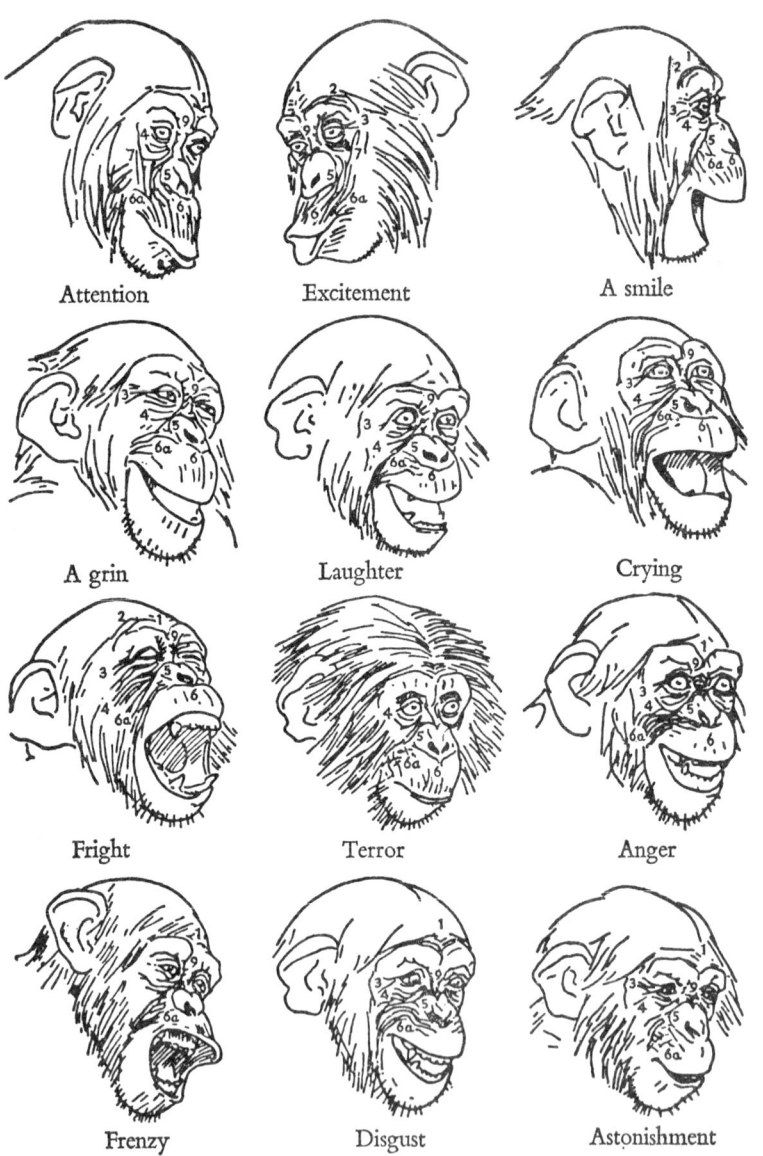

Attention	Excitement	A smile
A grin	Laughter	Crying
Fright	Terror	Anger
Frenzy	Disgust	Astonishment

FIGURE 6 Facial expressions of a young chimpanzee in various moods. Some of the creases are marked with numbers to emphasize that each is by no means confined to one expression (after Kohts).

(from Marler, *Darwin's Biological Work*, p. 170, 1959)

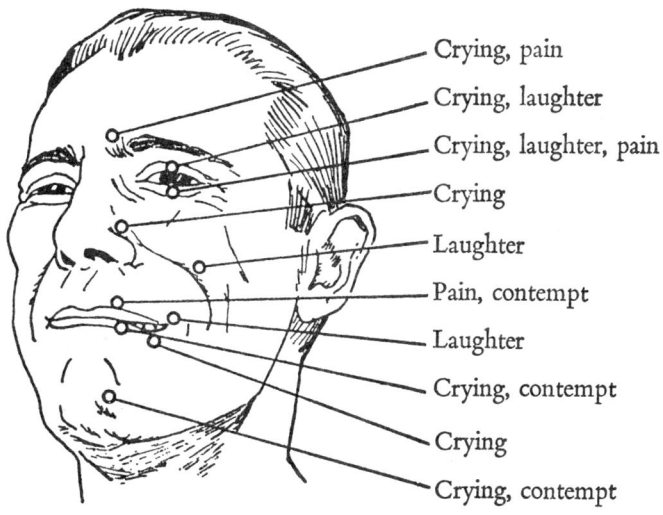

Crying, pain
Crying, laughter
Crying, laughter, pain
Crying
Laughter
Pain, contempt
Laughter
Crying, contempt
Crying
Crying, contempt

FIGURE 7 A diagram to illustrate that, just as in the chimpanzee, the same fold or wrinkle in the human face may be involved in several expressions (after Frois-Wittmann).

(from Marler, in *Darwin's Biological Work*, p. 171, 1959)

experience (Figs. 4 and 5). In the higher primates, particularly in the chimpanzee, facial expression becomes supremely important (Fig. 6), and there are many elements shared by several expressions just as is the case with our own facial expressions (Fig. 7). Amongst the higher felines and the wolves, as well as the higher primates, not only are we fairly certain that the meaning of most of the finer subtleties has to be learnt by experience; we can find excellent evidence that in many cases there is actual intention to communicate. So we arrive at evidence for a gesture language which

97

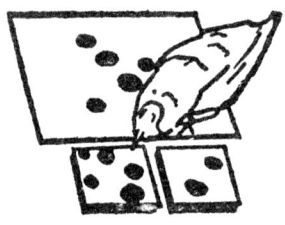

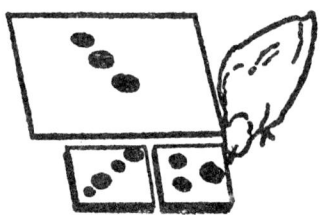

FIGURE 8 Jackdaw opening box according to a given 'key' number of dots on the big card. For further explanation see text (after Koehler, O., 1949).

is certainly at times intentional and which, it seems idle to deny, must often be consciously assessed by the recipient. How far is this different from human language? In answering this question we must[1] recognize a fact that is too often forgotten, namely that visual signals are as important and as complex in man as in any animal. And many domestic animals, even though they may have little power of understanding niceties of vocal communication, can nevertheless often recognize the intentions of their masters with uncanny precision – as was shown by the famous calculating horses that so puzzled comparative psychologists in the early years of this century.[2] What then is the difference, if any, between such visual signals and the language of human beings? It is not at first sight merely a question of symbolism because the visual signals of animals may become so reduced as to be little more than 'symbolic'. I have seen a bull European bison intimidate a junior and subordinate male in its herd by the very slightest sideways flick of its head indicative of the tossing and goring

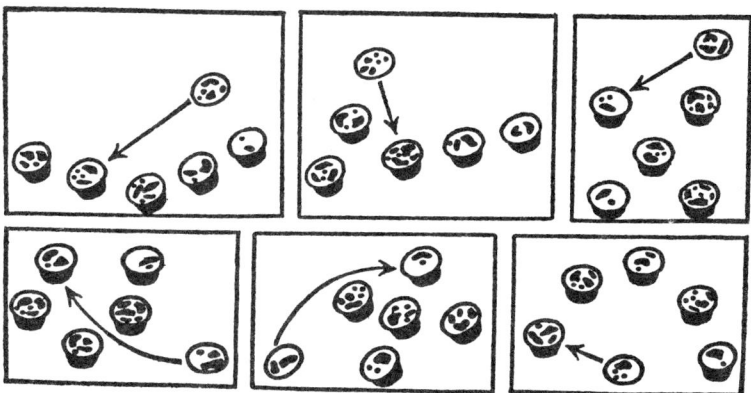

FIGURE 9 A problem successfully solved by the Raven 'Jacob'. The arrow points to each case from the 'key' pattern of plasticine pieces to the only box in each group of five which the bird opened. For further explanation see text.

(from Koehler, O., 1949)

movements. Nor is it, as used to be argued, that animal language is emotive only, whilst human language is supposed to be emotive *and* propositional. This distinction has been abolished by the discovery of the dance language of the honeybee. The dances are truly propositional in that they transmit precise information about the direction and distance of a food source – information which may be known only to the dancing bee until the dance is perceived by another bee and the information thus becomes transmitted. So this distinction likewise falls to the ground; but here it must be said that, amongst bees, we lack convincing evidence of any conscious intention to communicate.

We may ask, then, does the human ability to symbolize, in the sense of representing completely abstract or general ideas by words which in themselves have nothing of the essential characteristics of the concepts which they denote – does this provide us with a difference? For example, the word 'three' has nothing triple about it, nor is the word 'four' quadruple; but man can learn to understand absolutely general meanings of this kind: can animals do the same? As far as we yet know for certain, no animal language, however much information may be conveyed, involves the learnt realization of completely general abstractions. Otto Koehler and his pupils,[3] in their famous studies of the recognition of number, showed that animals, especially birds, can 'think unnamed numbers' – that is, they have a prelinguistic number sense; to some extent they think without words. But perhaps some examples of their number-training do involve something more than a special association between given numbers and particular signals; and there is now some evidence for the ability to arrive at a general solution indicating the comprehension of a numerical series.

Lögler[4] carried the investigation of the problem of number sense in animals a step further by his extremely thorough and painstaking work on counting in the Grey Parrot. Extending the earlier investigation of Koehler and his pupils, he found that a parrot 'Jako' was able to recognize the successive presentation of a number of optic stimuli as a signal for a task of performing the same number of actions. The bird, having been shown, say, four or six or seven light flashes, was then able to take four or six or seven (as the case might be) of irregularly distributed baits out of a row of food trays. Not even numerous random changes in the temporal sequence of signal stimuli impaired the percentage of correct solutions. Having learnt this task, a signal of successive light flashes was replaced by successive notes of a flute. The bird was, however, able to substitute immediately, without further training, and the change from light flashes to flute notes had no effect on the number of correct solutions. Nor was the accomplishment hindered by the completely a-rhythmic presentation of stimuli, or by a change of pitch. Although this parrot was not able to accomplish a task which represented a combination of the two faculties of learning numbers presented successively and simultaneously – e.g. he could not respond to numbers presented visually and simultaneously after hearing the same number of acoustic stimuli presented successively – yet when he had learned to 'act upon' 2 or 1, after hearing two sounds simultaneously or a single sound, he was spontaneously able to open a lid with 2 spots on it or a lid with 1 spot according to the same acoustic signals. That is to say, he was able to transpose from the simultaneous-successive combination to the simultaneous-simultaneous in twenty experiments without re-learning. It seems then that this remarkable work does bring our estimate of the counting achievements of birds a step nearer that of man; though it is still not true counting in the fully human sense.[5] (There has been some question as to whether the bird used by Lögler was truly naïve in the experimental sense, so even this most laborious study does not perhaps carry final conviction. But if it is repeated with the same result then we shall have a most impressive demonstration of the symbolizing ability of the bird mentality.) Perhaps the most reasonable assumption at present is [5] that however great the gulf which divides animal communication systems from human language, there is no single characteristic which can be used as an infallible criterion for distinguishing between birds and men in this respect. Human speech is unique only in the way in which it combines and extends attributes which, in themselves, are not peculiar to man, but are found also in more than one group of animals. We have evidence that animals can use conceptual symbols, but to a limited degree; and that here, as in so many other instances, the difference between the mind of animals and men seems to be one of degree – often the degree of abstraction that can be achieved – rather than one of kind. But man can manipulate abstract symbols to an extent far in excess of any animal, and that is the difference between bird 'counting' and our mathematics. I think we can sum up this matter by saying that although no animal appears to have a language which is propositional, syntactic and at the same time clearly expressive of intention, yet all these features can be found separately (to at least some degree) in the animal kingdom. Consequently the distinction between man and the animals, on the ground that only the former possesses true language, seems far less satisfactory and logically defensible than it once did.

The great advantage of airborne or water-borne vibrations, namely sounds, as symbols (whether for animals or men) is that sound carries far and fast; it readily by-passes obstacles, and there is a great spectrum of frequency ('pitch') and intensity ('loudness') available for use. Moreover, it is extremely economical; requiring a minimum of muscular effort to produce. So language, if it is to achieve its full potentialities for communication, must be a language of sounds. The language of sounds has been developed by the insects, the amphibia, the birds and the mammals. Perhaps the same is true of the fishes; this would indeed be expected, because a sound-wave travels farther, for a given loss of amplitude, in water than in air and also approximately four times as fast. But we have yet almost everything to learn about the significance of sound production by fishes.

Much vocal communication in animals is something very different in mode of development from human speech. A great deal of animal 'language' depends upon an innate ability to produce the required sounds or make the appropriate actions and a similarly inherent ability to respond by means of inborn mechanisms. But in man it is different; and there is no doubt that it is the flexibility of the technique of learned sound symbols and their manipulation, which constitutes the language of man, which has enabled man to progress so fast and far in the development and utilization of the basic forms of symbolic abstraction which he shares with the higher animals. That is why the communication of ideas is here linked with the development of ethical concepts.

There have been many theories as to the origin of human language; none entirely satisfactory, but probably most of them containing some truth. One of the latest[6] argues that the movements and sounds most important to a social animal are those made by its fellows; and that the physiological relation between the use of the larynx and the forceful use of the arms is one that provides still further evidence. An effort of the arms is accompanied by a closure of the glottis and a consequent temporary stoppage of vocal sound; and when the tense vocal cords are suddenly relaxed, there tends to be a gasp of released breath. This suggests that, in the earliest language, a loud 'bah', 'dah', 'mah', 'pah', etc., meant maximum human effort. These sounds are the most widespread and fundamental in the most primitive languages which have been described.

But how came it about that the ancestors of man developed their powers of abstraction to such a pitch that they were able to entertain these highly abstract 'ideas' derived from their experience of the world? They must have had ideas before they could produce a language adequate to express them. Before we can get much further with any attempts to answer this question we must give some thought to the physiology of perception and to the idea of a real world. We must ask ourselves how, in *our own* experience, we come to distinguish reality from hallucination, fact from illusion.

One of the most characteristic attributes of the scientific mind, and of the scientist's approach to his studies, is the common-sense conviction that he is dealing with a real external world. This was easy to understand during the last century, at the time when the physicists' concept of matter as composed of innumerable minute hard atoms resembling billiard balls, was not in dispute. Although many Victorian scientists were, no doubt, puzzled about the problem of knowledge; it was at least comforting that their studies seemed to reveal, in a sub-microscopic world, a landscape not different in kind from that perceived by the unaided sense organs. It is a remarkable fact that in spite of all the complexities of the modern scientific world picture; in spite of the immensely increased understanding of the nature of the nerve impulses as an ionic flux sending a propagated disturbance along a tube; in spite of the puzzling results of physiological studies of sense perception and of the action of the brain, which behaves as if it were a machine designed to deal with abstract symbols and nothing more – in spite of all this, the scientist still seems obstinately convinced that he is learning about a real external world. The difficulty with physics is perhaps fundamental. There is a pleasant story of Nernst in Berlin who, together with Max Planck, was, at the turn of the century, one of the dominant figures in the world of physics. They got in Einstein from Switzerland to join their team. Nernst in later years complained wryly, 'We hired Einstein from Switzerland as we might hire a lab boy; then the moment out backs were turned he messed up the whole of physics.' But it is true, as Eddington pointed out, that in one way even Newtonian mechanics constitutes a vast tautology. The essential advantage of concepts such as mass, momentum and energy lies in the fact that they lend themselves to mathematical treatment. And Heisenberg has remarked that mathematical formulae no longer portray nature; but rather our knowledge of nature. Yet somehow physical theories create, as it were from nothing, a series of definitions and concepts by the aid of which things, previously intractable, are brought together into order and intelligibility.[7] And if the scientist goes to the philosopher for help he will receive little encouragement. The process of induction is seen as little short of a scandal; there seems to be no straightforward answer to a sceptic's challenges, and just when an answer has reached the stage of being no longer questionable it becomes a tautology and therefore cannot serve as justification for any claim to knowledge. Yet *still* the scientist is convinced!

I have neither the ability nor the concern to follow this particular problem much farther; but the brief allusion I have made to it serves as an introduction to the consideration, from a biological point of view, of the question of the kind of concepts which the communication system of animals and men have to cope with. From this we can lead on to a discussion of some aspects of the biology of ethics and the evolution of morals.

In considering the probable origin of abstract ideas as a step towards trying to understand the development of the processes of communication in animals and men, we have to ask ourselves how it was that animals came to arrive at the perception of an enduring external world. Then, having achieved a germ of an appreciation of 'reality', how did they come to distinguish this from illusion and hallucination? I think it is fairly clear that the ability to recognize some external object as continuing to exist physically when not in sight must be practically confined to the vertebrates. I think most biologists would hesitate to state that this ability is consistently present in any except perhaps a very few outstandingly highly-developed invertebrates. Lower animals on the whole react to the stimulus of the moment only. They show little evidence of awareness of the continued coherence and existence of an external world. Even animals as high as birds and mammals will swiftly change their behaviour when under the influence of rapidly changing motivational states.[8,9] Under the influence of the maternal drive an animal such as a hamster will react to an offspring as something to be retained in the nest and suckled.[8] A moment later, under a different drive (if it gets outside the nest), she may take it as something to be eaten; showing no evidence whatever of a persistent consciousness of the reality of the object itself. But then the human baby shows exactly the same phenomenon, and it is only after some weeks of practice during which it has first begun to acquire perception of a three-dimensional world, that it 'realizes' that the missing rattle, for instance, is still there when hidden behind the fold of a blanket. I imagine that the animal's recognition of the continuance of an external world developed, at least in the mammals and birds, *pari passu* with the ability to solve new problems by conscious insight. That is to say, as soon as the animal was in a position to produce a new solution to an unfamiliar problem, without trial-and-error behaviour, but by a process of making a mental comparison and selection between possible alternatives based upon the relations which had been perceived – as soon as it could do this, then it could begin to see the existence and coherence of a real world. So I think that, in animals as in men, the appreciation of reality, as distinct from the rest of experience, must be through its coherence and lawfulness.

Sometimes one can see very clearly in animals the momentary confusion caused by mistaking an illusion for the reality. A Black-headed Gull attacking its reflection in a mirror, as so many species will, may stop the attack instantly the moment it is no longer opposite the mirror and so the apparition has disappeared.[10] Such an action gives the impression that the behaviour is a simple reaction only (as with a robin attacking a tuft of red feathers on a wire in lieu of a real robin).[11] A response of that kind is usually behaviour due to an inborn releaser. The gull does not go and look behind the mirror to see if its rival is still there. But some birds, and many mammals, *will* do just this; showing that they 'expect' the enemy to be round at the back. A really 'intelligent' animal, that is to say the one with the more highly organized perception of the world, will look behind the mirror perhaps once or twice and then, as I have myself observed in dogs, give up all interest in reflections for the rest of its life. I have, again in dogs, observed the same kind of response to pictures. A bitch of mine which growled anxiously on first encountering (in the year 1951) an oil-painting of a man leaning on his arms and looking out from the canvas, almost immediately 'realized her mistake'. Never afterwards did she give a moment's thought to the picture – either in the subsequent two or three weeks when she saw it daily nor when she has since seen it; which she has done repeatedly, at intervals of a year or two, to the present time. And so it seems that the first step in the process of achieving awareness of reality must have arisen, probably a number of times, both low in the vertebrate scale and probably high in the invertebrate scale too. Another line of evidence comes from the study of play. In the mammals strikingly, and to a much lesser extent in birds, one sees many examples of play in which one feels inclined to say without hesitation that the animal 'knows' that none of it is 'in earnest'. One gets the overwhelming impression of a world of arbitrary rule and make believe. If this interpretation is correct, then again we have the potentiality of distinguishing between the worlds of fact and of fantasy. Whether the occurrence of dreams

106

FIGURE 10 *General Caption.* Evidence for conclusion that a bird does some 'inward marking' of the units he is acting upon is provided by occasional 'externalization' of these supposed inward marks in the form of intention movements. Thus a Jackdaw, given the task to raise lids until five baits had been secured (which in this case were distributed in the first five boxes in the order 1, 2, 1, -, 1) went home to its cage after having opened only the first three lids and having consequently eaten only four baits. The experimenter was just about to record 'one too few, incorrect solution' when the Jackdaw come back to the line of boxes. Then the bird went through a most remarkable performance: it bowed its head once before the first box it had emptied, made two bows in front of the second box, one before the third, then went farther along the line, opened the fourth lid (no bait) and the fifth and took out the last (fifth) bait. Having done this, it left the rest of the line of boxes untouched and went home with an air of finality. This 'intention' bowing, repeated the same number of times before each open box as on the first occasion when it had found baits in them, seems to prove that the bird remembered its previous actions. It looks as if after its first departure, it became aware that it had not finished the task, so it came back and started again, 'picking up' *in vacuo*, with intention movements, baits it had already actually picked up; when, however, it came to the last two boxes which by mistake it had omitted to open on its first trip, it performed the full movements and thus completed its task.

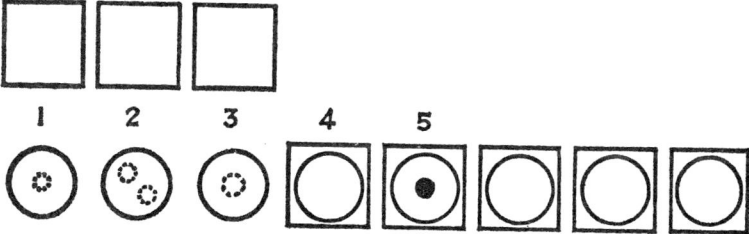

FIGURE 10 *Special Caption* The Jackdaw has removed the cards covering the first three boxes and has taken the four baits which they contained. The remaining boxes are closed; No. 4 is empty and No. 5 contains the last 'allowed' bait
(after Koehler, O., 1949).

FIGURES 11 and 12 *General Caption* As a result of recent developments in the technique of brain stimulation for the study of the innate behaviour of the domestic fowl, E. von Holst and U. von St Paul (1960) have shown that point stimulation by oscillating current at a wide variety of loci in the brain-stem can elicit a broad range of both simple and complex movements belonging to many different instincts. Among the most characteristic of such instinctive behaviour patterns that can be elicited electrically in this manner is behaviour appropriate to attack on or to flight from different kinds of enemies. Thus the behaviour appropriate when threatened by a hawk in the sky is very different from that appropriate when threatened by a weasel on the ground. Both these complex patterns of behaviour can be elicited by electrical stimulation, in particular parts of the brain stem. Fig. 11 illustrates a particularly interesting example of the combination of electrical stimulation in the appropriate place, together with presentation of a stuffed polecat. The fowl was a naïve animal in the sense that it had never before, as far as is known, experienced the attacks of a polecat. The electrodes are extremely minute and are attached in such a manner as to leave the bird complete freedom of movement. The figure shows that when the bird is not stimulated electrically, it reacts in a partially aggressive manner to the appearance of the stuffed weasel. If, however, while it is doing this the current is turned on so as to stimulate the brain mechanism responsible for controlling the animal's response to dangers of this kind, its activities are enormously increased and it launches out into full-scale attack upon the dummy. Now a live, wild polecat or weasel that does not retreat when attacked in this manner is a very dangerous animal indeed to a fowl, and the natural consequence of such an encounter with an animal that stood its ground would be ignominious retreat. This retreat can be brought about even in response to a stuffed polecat as the result of electrical stimulation. But even more remarkable is the fact (Fig. 12) that a higher intensity of stimulation in the same place will, even in the absence of any stuffed dummy, cause the fowl to behave as if it were attacking a weasel though there is no weasel there at all. In this case the fowl has every appearance of being the victim of a hallucination, for it flaps away from nothing, screaming with fright.

FIGURE 11 *Special Caption* Releasing ground-enemy behaviour. Without a suitable object the stimulated hen shows only locomotory unrest. Towards a fist she shows only slight threatening A. A stuffed, motionless polecat is vigorously threatened and attacked; if the stimulus ends at this moment, the hen remains standing and threatening slightly B; if it does not end, she checks and flees, screeching C (after ciné film records). (von Holst, E. and von St Paul, U. 1963)

FIGURE 12 *Special Caption* A Behaviour sequence of 'fleeing from a ground enemy' in the absence of any stimulus object but with slowly increasing stimulation of the field (or constant stimulation of moderate intensity); B the reaction to a sudden strong stimulus. (von Holst, E. and von St Paul, U. 1963)

has been another aid to achieving this most vital of mental steps, is a difficult point. Watching dreaming dogs and cats one certainly gets the impression that the animal is undergoing a vivid mental experience. But it is hard to feel sure that anything of this is carried over into the waking world. Again and again I have watched dogs wake up from dreams and never have I seen any suggestion that the emotional effect of the dream is carried over, even momentarily, into real life. The same seems to be true of cats. Perhaps it may not be long before we achieve an electrophysiological technique for comparing the disturbances in the centres of consciousness during dreams with those which occur during wakefulness; determining whether or not the effect of the former is carried over into the latter. Not until some such method of approach is possible shall we, I think, get much further in our understanding of animals' dreams.

The question of hallucination is of course a much more subtle one, which we can hardly begin to consider in relation to animals. All one can say is that it is possible to induce what *appear* to be hallucinatory experiences by electrical stimulation of the brain in animals as in men. Thus, if an appropriate focus in the brain-stem of a domestic fowl is electrically stimulated, actions appropriate to attack on, and, at higher intensity, flight from an enemy are facilitated. This effect can be coupled with the effect of a visual stimulus. Thus the sight of a human hand alone may cause a minimal response; but if electrical stimulation is applied at the same time the bird's response is much bolder and may pass from threat to actual attack. Similarly, if the stimulus is a stuffed weasel or polecat – in any case an object which is both more powerful in stimulating attack and more frightening – the response may quickly be increased by the electrical brain-stem stimulation to the point at which the stuffed weasel 'wins' and the fowl flies in screaming retreat. Finally more intense electrical stimulation may, by itself, cause the bird to flee from a spot where there is nothing at all – *as if* the bird had an hallucination of some terrifying enemy at that spot. (For fuller explanation see Figs. 10 and 11 with their legend.)

The recognition, whether by animals or men, that there is an external world of consistency, while undoubtedly a tremendous advance, leaves still an enormous gap to bridge; namely the gap between the practical, real world and the way to behave in it; and the concept of some deeper, more consistent reality behind it. Not until the mind can entertain some vision, however fleeting and fragmentary, in which things are as they ought to be; in which duties and ideals demand allegiance, a world in which ethics and morals apply – not until this basic idea of metaphysics is entertained, can the lineage leading to man have taken the first steps into the domain of real manhood.

MAGIC AND RELIGION

It is when we come to this distinction that we begin to consider the relationship between the development of magic and the beginnings of true religious awareness. As to the first, what looks like magical behaviour can arise very early and in a very simple fashion. It is called, by comparative psychologists, superstitious behaviour. A single example will suffice. One of the most striking as well as the most amusing examples of this superstitious behaviour is told by Konrad Lorenz[12] who had a pet Greylag Goose (*Anser anser*) known as Martina, hand-reared from the egg and fully imprinted on him. Martina slept in his bedroom, flying out of the window soon after dawn to spend the day in the garden. She was not, however, capable of flying back into the room by the window in the evening but had to come in at the front door and walk upstairs to bed. The centre of Lorenz's house at Altenberg is occupied by a big hall with a gallery running round it at first-floor level. The first floor is reached by a staircase which curves up on the right-hand side of the hall as you come in from the front door. Opposite the front door at the other end of the hall is a large window. The goose had been trained from her earliest days to come in when the door was opened for her at dusk, to walk upstairs to bed. At the very beginning of her training, just as she came in at the door, she took fright; and, naturally, she fled towards the window at the other end of the hall and had to be comforted and coaxed back before she could go upstairs. On many occasions after this occurrence she would make a panic-stricken dash towards the window before turning and making her way upstairs. With the passage of months, this 'detour' was gradually eliminated until, after a year had elapsed, nothing was left of it save that instead of taking the shortest route from the door to the nearest point of the bottom stair, she would march straight towards the window until opposite the *middle* of the bottom stair; then do a sharp right turn and, the 'ritual' completed, walk upstairs. One day the goose was forgotten and left in the garden in the dark until long after her normal bedtime. When she was remembered, the door was immediately opened and a very flustered and agitated goose came in, obviously in a great hurry to go to bed. She darted directly to the nearest point of the bottom stair, ascended five steps, and then stopped suddenly as if in terror, stretching her neck and spreading her wings. She then gave the alarm call, came within a hair's breadth of taking wing, but stopped herself in time. Instead she turned and climbed hastily down the five steps to the floor, and proceeded to her usual place opposite the middle of the bottom stair, facing the window, did a sharp right turn and started upstairs again in the normal manner. On reaching step No. 5 she stopped, looked around; then she shook her feathers and 'greeted' – both actions normally seen in gray geese calming down after experiencing a threat of danger. It was clear that the ritual had become a necessity to Martina, the omission of which gave rise to feelings of anxiety and fear. So what appears to be magical behaviour could have arisen without any metaphysic; without any concept of absolute values of any kind.

Magic of this kind was, I incline to think, the beginning of utilitarian science rather than religion. It could easily have been the start of the attempt to control nature; it is to me much less easy to imagine how, except at a much later stage, when science had become a search for understanding for its own sake, it could have led on to religious awareness. Religious awareness, it seems to me, partakes much more of an intuitive realization of participation in, or belonging to, a system vastly greater than oneself. This is nearer the approach of the artist and recalled Benedetto Croce's ideas as to intuition; so luminously discussed by Joyce Cary[13] According to this it is only by intuition that we have direct knowledge of the world. How did this intuition, this direct knowledge of a universe infinitely greater than, but in some sense sympathetic to ourselves, arise? For until it did arise, in embryonic form, I see no possibility of advance to the next stage.

We can hardly begin to discuss the nature of primitive ethical concepts without speculating about their possible origin and course of development. This is a path along which pitfalls of every kind abound; as is I think shown only too clearly by many previous speculations on the subject; even those emanating from the most learned. It seems safe to say that we shall never know more than a tiny fraction of the facts which would be necessary to arrive at anything like a coherent and convincing account. Nevertheless, my justification for adding a few further speculations of my own is that I feel that the development of the study of animal behaviour in recent years has given us some new understanding bearing on the subject.

The evidence that some quality of early man, call it if you like 'spirituality', did in fact exist is overwhelming. This 'spirituality' is a quality which may manifest itself in animism, worship of the dead, and other similar characteristics of the human stock. It is when we come to try and disentangle these different manifestations and place them in evolutionary sequence that the difficulties thicken about us. First as to their antiquity. Magic and spirituality certainly antedate the appearance of *Homo sapiens* in Europe. Neanderthal man had some kind of a skull cult and also funeral rites. It has even been suggested that Pekin man, *Homo erectus pekingensis*, may have engaged in ritual cannibalism; for here again innumerable skulls have been found with broken occipital parts – as if for extraction of the brain for ceremonial consumption. But I think the conclusions of modern ethology should serve to make us careful in our attempts to relate these various manifestations. I agree with Blanc[14] when he says 'the main object of the ethnologists and historians of religion has often been to resolve the ever-apparent complexity of beliefs into categories that existed only in their own minds'. He goes on to say that probably none of the categories that these investigators, such at Tylor and Fr Schmidt, employ has ever been evinced anywhere in the pure and simple form in which they were conceived. They have always appeared associated in various proportions with other beliefs, pertaining to different categories. The error of authors in the past has, so he argues, been the attempt to construct artificial and arbitrary evolutionary series. This done, the assumption is made that their categories once had a real existence as pure and isolated ideologies.[15] The point I wish to make here is that certain of these ritualistic practices (e.g. magic) could have arisen long before there was anything deserving the name of spirituality. By the ordinary principles of conditioning it is possible for an animal to develop, from the kind of superstitious behaviour described above, actions which appear for all the world (as in Konrad Lorenz's goose going up the stairs to bed) to be a rite designed to placate some malevolent spirit. Once then the development of man as a species, having a long childhood and closely-knit social structure, had occurred, there was a possibility that imitative tendencies and social cohesion might cause such examples of propitiatory magic to be passed on from generation to generation for a very long time. So I think it is easily possible to imagine magical practices of a simple kind antedating any theory of the existence of ghosts or spirits. It is highly probable (see p. 74 above) that the earliest religious ideas arose through the contemplation of the dead and the fact of death being associated with dark caves. But even here observations on animal behaviour give one pause. There are instances of animal behaviour which suggest grief and mourning. (At least one badger watcher has described the distraught cries and apparently anguished behaviour of a 'funeral', and African elephants have been seen apparently attempting to revive, or remove a dead associate) [16]. This makes us wonder whether some sort of funeral rites could not have arisen much earlier than we previously imagined. It may well be that the pathetic pining and self-immolatory attachment of a sheep-dog to his dead master (such a contrast to the distressing attacks of dogs upon their masters referred to in note [9]) may be the expression in the world of the domestic animal of some profound characteristic which goes far back into the history of the wolf pack. So if we are to make distinctions it seems plausible to argue that the development of something like superstitious magic into a desire to get power over nature, was the forerunner of the scientific attitude; and that the developing awareness of death was the forerunner of the religious attitude. But even though these two attitudes have arisen separately they must have crossed and recrossed in evolution, fusing and separating again in a mysterious network. So it could happen that, on the one hand, religion became associated with magic and superstition, whilst on the other, the desire to control the natural world became associated in modern science with an intense desire to understand it (derived itself from religion), to acquire knowledge about it for its own sake. But the primary distinction between magic leading to science on the one hand, and worship leading to religion on the other, is to my mind fundamental. I have no sympathy with the assertion that the question 'What is true?' can be whittled down to the question 'What is useful?' I agree with Dr Habgood [17] when he says 'apologists for science often claim that the one sufficient criterion of a good hypothesis is its ability to make successful predictions. But this is not so. In the last resort it is only strictly true of technology.'

But whatever the course evolution may have taken, it seems

safe to assume that magic and religion (and also art as partaking of something of both), must have been powerful factors making for cohesion in early hominoid societies. Thus there is no difficulty in seeing how natural selection might have helped to foster them all. Once the first steps in the direction of supernatural beliefs and practices had been taken, it is not too difficult to understand how both magic and religion could have branched out and flowered in the multitudinous forms which we know. Perhaps it is more remarkable that, as was stressed in the last chapter, many of these cults should have been so persistent, so uniform, as to be recognizable in fragmentary remains over so many thousands of generations. What I have been suggesting as a result of thinking about modern studies of animal behaviour bears out, to some extent at least, the views of R. R. Marett.[18] I think it would be a mistake to suppose that in the *very earliest* stages of development of his beliefs (preceding the stage of 'spirituality' referred to above), primitive man made any firm distinction between matter and spirit. Perhaps the 'ghost soul' spoken of by Tylor should not, in its earliest stage, be thought of as either material or spiritual. At any rate, l am attracted by Marett's pre-animistic conception which attributed an ill-defined power of virtue to all things and evoked awe in the mind of primitive man. And that, as the idea of continued existence after death fused with that of the power of ritual in controlling nature, something very like the pre-animistic conceptions of Marett may have been reached. However this may be, once the idea of a continued personal existence, if not on an immaterial plane, at least not attached to the known physical body, had been reached, then the concept of the worth or value of such personality would be likely to follow; for who could not fail to be more impressed with the worth of one's fellow beings if one felt convinced that they, and oneself also, were partakers in some kind of eternal existence?[19] And so once the 'value' of human personality had entered the mind of man, the possibility of morals came with it.

Before coming on to a discussion of the natural history of

ethics I wish to allude briefly to one other problem which I think arises at this stage. Once the idea of the value of a personality – thinking of it as something worth while in itself – has been reached, then a comparison of the worth of different personalities becomes inevitable. And so we reach the stage at which the germ of the idea of mental normality or mental health comes upon the scene. And this is not merely an appreciation of power, a simple reverence or fear of the stronger or the more clever, but something far more subtle.

In the first instance, a being that is mentally abnormal tends to be seen as anti-social and dealt with as such. And this I think subsumes all the reactions to such beings in the most primitive societies that we can imagine. But as the concepts of magic and 'mana'[20] developed, mental maladies of particular kinds must have become most mysterious and awesome. And right to our own day we find the tendency to regard such afflictions as signs of possession by evil spirits. So, from being anti-social, some such individuals have come to be reckoned as particularly to be cherished. Psychologically, much mental disease involves, I think, some tendency to self-destruction; some tendency which results in a breakdown or fragmentation of the personality or of the totality of experience. The mentally impaired individual is partial instead of whole in ideas and ideals. The whole man is the sane man. This of course is not to say that the healthy man cannot be eccentric. Provided his ideals are of wholeness, he can be 'mad' about certain things – for example the 'madness' of Beethoven or of many artists. This lies at the very core of the development of civilization. Thoday has remarked[21] that 'biological progress not only involves increasing versatility of individual species but also increasing diversity of species *harmoniously adapted to one another*'. That I believe is a very profound summary of the present day view of evolution by natural selection as apparent to a geneticist. I think we can apply it with equal or greater force to human society. Progress in human society involves increasing versatility of individuals and increasing diversity of individuals *so long as they are harmoniously adapted to one another.* Dobzhansky[22] comes to this view in his quarrel with what he calls 'the implied assumption that equality of opportunity and mitigation or elimination of intergroup competition is tantamount to uniformization'. He argues that 'a large and complex society should be better able to provide for the specialized talent and to tolerate unconformity than a small homogeneous group. I for one do not lament the passing of social organizations that use the many as a manured soil in which to grow a few graceful flowers of refined culture.' Whether we agree with that sentence or not, I think we must agree that the development of tolerance of diversity and versatility within a social group is perhaps the most important single factor permitting and encouraging social progress.

ETHICS

For the moment I am going to leave aside the question as to whether one can successfully base any ethical principles upon evolutionary principles. The problem is beset with difficulties. Thus, for example, in nomadic life (e.g. the surviving bushman clans of the Kalahari) the duty of helping the weak and old may have to be pitted against the duty of preserving the group. In a strongly individualistic culture of the Victorian type, the claims of charity are compelling and obvious. But in our culture, which has reached the point at which the abolition of poverty by organized effort has become possible, individual charity might again be deprecated as weakening the drive to hasten the full emergence of the welfare state.[23] But if Thoday's view of the evolution of society is correct, then it would seem that any codes of behaviour which increased versatility of individuals and diversity of individuals within a harmoniously adapted whole would be ethical. Such a conclusion, however, needs qualification. While there are, I think, overwhelming arguments for assuming that there has been real evolutionary progress, and that this has led by a series of stages

118

to the most highly evolved type, namely man; yet principles which lead only and solely towards the further perfection of man, while ignoring completely his environment, may in the long run prove far from ethical. Man is a species living in extremely complex ecological inter-relationships – an ecological community in fact – with a great many other species. Ultimately he is dependent upon these for the continuance of his life. If he upsets that state of dynamic equilibrium which is popularly known as 'the balance of nature' too severely and too suddenly, disaster may supervene. The stability and resilience of the whole system is such that quite large perturbations can be absorbed and eventually smoothed out. But if the shocks are *too* violent a point of no return may be reached in which the changes in population of the animals and plants concerned may start to avalanche uncontrollably with disastrous results. As Waddington[24] remarks, it is as remarkable as it is unfortunate that biological theory in the field of population dynamics should be so backward. Synecology, as it is called, that is in effect the scientific and quantitative study of the web of life, is a part of biology which should form a basic element in all economic and political theories. Without this, as the human population grows, we approach daily nearer disaster. The industrial world seems simply to utilize as much as possible of any sources of energy that are available. The uncontrolled exploitation of the fertility of the grain-growing lands of the new world, producing the familiar dust-bowl effect, is a single, though rather extreme, example of what man tends to do to his environment. Most primitive societies are either small enough, or sufficiently well adapted to their environments (or both) to live in harmony with them. But with modern man it is not so, alas. Considered solely as an animal, man is just about the nastiest creature that has ever been evolved! He is nasty not primarily through lack of good principles (which abound – though there are more than enough bad principles noticeable as well), but because he is increasing too fast and because of his ignorance and stupidity where he at least has the means to be far-seeking and sensible. This seemingly savage criticism must not be taken to imply a whole-hearted condemnation; nor a self-righteous promulgation of edicts from the superior level of the scientific viewpoint. It is impossible that all branches of learning, all types of education, and all technological advances should keep pace with one another and come in the right order. All of us some of the time, and most of us all the time, have to take action upon inadequate evidence. One must not therefore blame those who have advanced hygiene and medicine, developed agriculture and promoted communications, brought peace and plenty where there have been war and famine, for the population explosion that has resulted. The true lesson to be drawn from all this is, I think, that which I have already stressed when talking about sociological and biological progress. It is essential to ensure that we cease tottering from one crisis to the next. To do this it is needful not only to love the good and our fellow man (which is of course basic) but also to ensure that all branches of science, arts and technology are harmoniously developed in the service of man as a physical and spiritual being; for if we neglect even one of them too long we may be heading for disaster. It should be only too obvious now that, faced with our present population explosion, we need not only control of the reproductive rate on a worldwide scale but also to provide better distribution of produce and a greater productiveness which does not at the same time threaten, as some of the new pesticides may be doing, the continued existence of the living environment which we inherit.[25]

All this, up to a point, is only common sense; but it is common sense which in my view needs restating again and again. And in many respects the scientist and the technologist should be in a position to offer leadership if only they take a sufficiently wide view of their social responsibility; for by the very nature of their studies they are often enabled to see just a very little further in certain important directions than are many of those who are trained in other fields.

But if what I have just written sounds too prosaic and utilitarian, I would redress the balance by quoting Robert Bridges once again:[26]

> . . . the high goal of our great endeavour
> is spiritual attainment, individual worth,
> at all cost to be sought and at all cost pursued,
> to be won at all cost and at all cost assured;
> not such material ease as might be attain'd for all
> by cheap production and distribution of common needs,
> wer all life level'd down to where the lowest can reach.

Let us now look at one or two other problems in which recent developments in biology may shed a little light on the path. I have already alluded to the need for population control: what can we say then about the ethics of eugenics and the possibility of selective breeding? Here I think Dobzhansky takes a sane and balanced view; and in what little I have to say I largely follow him. First it must be said that the view, put forward in some quarters, that the development of culture means that our genes are no longer important is quite without foundation. I have been arguing above that genetic changes in the human species must have been influenced – to my mind very greatly – by culture. But the reverse is undoubtedly true, although perhaps to a lesser extent. Gene changes do undoubtedly influence culture; the influence works both ways. So much nonsense has been talked about eugenics that the subject has become rather disreputable. This is unfortunate, because whatever man's future, we must continue to depend even more than in the past in the interaction between genetic make-up and social organization. Here the concept of the 'genetic load' is important. This 'load' includes not only genetic diseases manifesting themselves in gross malformations or in syndromes such as haemophilia, retinoblastoma or sickle-cell anaemia. Even more important in the long run may be more subtle effects such as lack of resistance to stress and infection, loss of powers of mental and physical adjustment, and so forth. All these add up to constitute the genetic load. Added to this there is the incalculable effect of exposure to radiation, drugs and various deleterious influences of industrial life in general. Where there is still so much to be learnt it is not surprising that geneticists take strongly opposing views as to the seriousness of our genetic plight. Some of the arguments of those such as Muller,[27] suggesting that we are entering the genetic twilight of the race and that deterioration has now gone too far to stop, rest at least in part on strong arguments. But before accepting the prophecy of doom we must remember that there are several pieces missing from the puzzle. We do not know what part of the genetic load consists of unconditionally deleterious variants, the supply of which is maintained by recurrent mutations. We do not really know whether the evolutionary processes now under way in human stock are tending towards biological deterioration or not. It does seem fairly clear that modern civilization has the general tendency of increasing the mutation rate in mankind. But even this does not tell us what we ought or ought not to do. For some geneticists like Muller and Dobzhansky, any increase in mutation rate, whether large or small, is undesirable. Others might consider it differently, and R. A. Fisher[28] at any rate used to argue that some increase in the rate of release of inherent genetic variability might well be a good thing for the human race. With such striking differences existing amongst geneticists on quite fundamental topics, it need hardly be said that, whatever the future may hold, genetical knowledge is at present far too restricted for eugenics to have any important part to play in our generation. But the future may well lie with it: time alone can tell.

Another aspect of this general problem is that of the class structure of society. Is the welfare state going to be a good or a bad thing for us? Richter[29] is one of the pessimists. He foresees the biological twilight of human evolution brought about by the influences of culture. Our ancestors, he says, lived the wholesome life of wild animals. They struggled and fought for survival; all but the strongest and cleverest succumbed. It was this that resulted in the development of man's finest qualities. Now, with the necessities of life assured to everybody, with the environmental hazards and epidemics controlled, the unfit survive and reproduce their kind. The process that has transformed the wild rat into the domesticated laboratory rat is also working in human evolution. To this it can be replied that it is true the laboratory stock is different from wild stock. But it does not follow from this that they are decadent and unfit; nor does it follow that the welfare state is having the same effect on mankind. What Richter has overlooked is the fact,[30] and an obvious one at that, that the laboratory rat is manifestly fit to live in its environment – which is a laboratory cage. It is not a decadent product of the absence of natural selection; it is a product of rigid selection in the laboratory environment. It is tame and tractable, unaggressive and fecund; which amounts in fact to a high degree of fitness. It is also fit in that these qualities are associated with lessened activity of the adrenal and thyroid glands so that this state too contributes to fitness. Therefore it does not follow necessarily that either civilization or the welfare state will set aside natural selection; what they do, however, is to change its emphasis.

Reference was made above to the possibly great future rôle of eugenics when it comes to be based, as we hope it will, on foundations of knowledge incomparably greater than at present. With the development of artificial insemination in human beings, the picture is changing very rapidly. The time may not be far distant when human semen will be preserved indefinitely in the frozen state without deterioration so that, as Muller suggests,[31] we could at will populate the world with Einsteins or Leonardos, Lenins or Lincolns, Pasteurs or Picassos. And if techniques are developed to make it possible to implant nuclei of somatic cells into enucleated eggs, and then force them to develop parthenogenetically, we could even get rid of gene recombination which takes place when the germ cells are formed which causes so many of the progeny of outstanding parents to fall short of the parental standards. By such methods it should be possible to produce *exact* replicas of the geniuses of the past, and in any number. To

123

quote Muller: 'the biological distance from apes to men is a relatively slight one, yet how potent. Our imaginations are woefully limited if we cannot see that, genetically as well as culturally, we have by our recent turning of an evolutionary corner set our feet on a road that stretches far out before us into the hazy distance.' This indeed is a startling prospect, and certainly no one can say it is impossible of realization. However, I feel some sympathy with Dobzhansky when he retorts, 'Are we, hastily made over apes, ready to agree what the ideal man ought to be like? Muller's implied assumption that there is or can be *the* ideal human genotype which it would be desirable to bestow upon everybody is not only unappealing but almost certainly wrong – it is human diversity that has acted as a creative leaven in the past and will so act in the future.'

But to descend from these dazzling heights to more personal and immediate problems, has the biologist anything useful to say about such present day worries which the terms chastity, monogamy and homosexuality – to mention only a few – raise in our minds?

As to monogamy, the story of the animal kingdom provides but little help. Among certain mammals, for example wolves, and among many birds there seems to be lifelong monogamous marriage which, at any rate in wolves, is associated with intense personal affection. With the primates, monogamy occurs but seems to be the exception. Thus tarsiers (*Tarsius*) are said to be monogamous, the pair nearly always being found together, often with the youngest offspring accompanying them. The Lar Gibbon (*Hylobates lar*) also has a mating pattern resembling that of human beings. The sexual drive of the male appears to be low, and the animals live in groups consisting of one male, one female, and their young. But apart from these not very convincing exceptions, the primates appear to be a polygamous crowd. In considering this problem, however, it must be remembered that the reproductive biology of man is a rather unusual one amongst mammals. The general rule in mammals (there are many exceptions) is for the female to be receptive to the advances of the male only at certain more or less limited stages of her oestrous cycle. This corresponds to the time at which fertilization is most probable and occurs usually at some particular time of the year. However, in many primates the oestrous cycles of different females are not in step so there will usually be, at any one time, one or more receptive females present. The human female differs from all in remaining more or less uninterruptedly receptive throughout her reproductive life; and this (as Dobzhansky shows) favours the development of the monogamous family. Now males tend to be sexually aggressive and indiscriminately ready to mate with any and all receptive females; whereas the females in wide ranges of the animal kingdom are coy, and their reluctance has to be overcome. This does not apply to species that are strictly monogamous, where the mates are permanently paired. And it can be shown that natural selection, other things being equal, favours sexually aggressive polygamous males and passive discriminating females. The energies of a non-human primate male are largely devoted to keeping out competitors and protecting the band from external enemies. He does little else in the community; and in return for this service gets first choice in feeding and first choice at females; so he can always go to the most sexually receptive one on any occasion. If this was the organization of the ancestors of man, a big change must have taken place with the advent of tool-making ·and the adoption of an upright posture appropriate for ground living. Continuous sexual receptivity, it can be supposed, now made monogamous family life possible, for it reduced the number of unattached rivals to a minimum. The male could thus now specialize in hunting and food gathering leaving the female at home to do the chores and in particular care for the fire. Aggressive action against individuals was now less necessary; domination had changed to co-operation, not only with the mate, but even with other males who could now co-operate in the process of hunting. This co-operation in turn made communication a necessity, and so the development of language was further stimulated. So we seem to see plausible reasons for taking the view that monogamy is the natural state of man. No biologist, qua biologist, would I think go so far as to argue that 'chastity' (in the sense of marital fidelity) was the greatest of virtues. I do not want here to enter this heavily mined area except to say that, in marriage, 'chastity' (in this sense) and charity are seldom incompatible. Where, in all honesty they are really found to be so, it cannot be doubted that the latter is more important. In this matter, then, 1 incline strongly to the attitude expressed recently by the eleven Quakers in their *Towards a Quaker View of Sex* (1963). I like too in this matter to sit at the feet of no less a person than Dean Inge who said that if Christ were living today he might well say that marriage was made for man and not man for marriage.

But as to the problem of pre-marital intercourse, I feel we are as yet far too ignorant of what is really happening and what the effects are. I believe this is one of the fields in which honest inquiry together with unprejudiced yet loving concern, such as the churches should be in the best position to give, is of the greatest importance. And I am encouraged that at least some Anglicans and some Quakers see this as a prime duty. I like Macmurray's definition of chastity as simply 'emotional sincerity'. In the majority of cases, pre-marital intercourse (at least where promiscuous) *appears* plain selfishness, disguised, in self-deception, as something worthy, or at least excusable. It is often called 'love' (which is unselfish giving) when in fact it is the reverse – selfish taking. If Macmurray's dictum were widely believed and understood there would be less trouble. One may well feel that any liberty or concession which leads to the production of even a single unwanted or uncared-for child is unlawful for the Christian and humanist alike – for in Christian culture today the unwanted child carries a potential load of social maladjustments which weighs not only on the child but constitutes a major problem for society. But even here we know too little. This conclusion as to the handicaps of unwanted children would certainly not apply in many non-Christian and non-European cultures. Even in Scandinavia the ill effect seems to be little in evidence; although some of the maladjustment (e.g. high suicide rate) usually attributed to other causes, may in fact be linked with it.

There is another, in comparison small but none the less troublesome ethical problem – that of homosexuality. According to Westermarck, the severe condemnation of homosexual practices in Christian countries is to be traced to the abhorrence felt for these acts by the ancient Hebrews on account of their association with idolatrous cults.[32] Here and now we have laws so savagely vindictive that it is almost inconceivable that anyone today can be found to defend them. That there is any widespread resistance to change is all the more extraordinary when it is realized that Parliament in 1885 created the offence under which most of the prosecutions take place, namely gross indecency, entirely by accident and unwittingly. Here I think the life sciences provide us with much stronger arguments than they do over the question of monogamy and chastity. First one can say without any fear of contradiction that there is not one iota of biological or psychological evidence to warrant the labelling of homosexuality as 'unnatural' in the young. On the contrary, it seems to be natural right through primitive society, widely through the lower mammals, and certainly not infrequent in the socially more highly organized of the birds. In innumerable species young animals are at first unable to distinguish male from female and attempt to copulate indiscriminately with either. In many highly social groups, from the fishes upwards, males and females frequently adopt the behaviour of the opposite sex; alternation in this respect being linked with a temporary state of dominance or submissiveness in which the individual happens to be. Permanent adult homosexual partnerships are known in creatures as far apart as geese, porpoises, monkeys, elephants and giraffes. Although, however, we can state homosexual behaviour to be natural, that is only part of the answer. And clearly the relationship could, if widespread, become highly dysgenic. I personally think we can safely allow selection, even under the conditions of civilized life, to take care of this problem for us. I may be wrong, but I cannot conceive a situation – nature being what it is – in which the structure and stability of society would be shaken, let alone gravely threatened, by homosexuality.

There remains one other ethical problem [33] of a stature so huge and of an urgency so imperative that it seems to reduce all these others to mere trivia: I refer of course to the problem of aggression and war. This is to me *the* moral problem of this age; and I confess to a feeling of outraged fury when I find spokesmen and authorities in the Christian churches allowing problems of, say, church order, to absorb their energies out of all proportion to their importance – with the result, so it seems to me, that this gravest of issues takes a low place or is ignored.

I think here the science of ethology has its mite to add to consideration of the problem. Earlier I gave an account of Lorenz's views as to important differences between the anonymous swarm and the primitive forms of family life such as the clan or group which is characteristic of the higher mammals. It can be argued, as indeed Lorenz does, with a good deal of plausibility, that man cannot manage any of his instincts simply because he has modified his environment so fast that none of them fit his new circumstances. The first task for man, then, is to render harmless the new unoriented and intraspecific aggression; and it is characteristic of aggression, through large areas of the animal kingdom, that it is more easily content with substitute objectives than are many other instincts. Hence the classical concept of catharsis. Accordingly Lorenz would argue that among other activities, international competitions, especially dangerous international sports, are an important factor in amelioration of a more dangerous social aggression. He would extend this view to assess the possible advantages of space flight as a means of redirecting the aggressive tendencies of large groups of people. The great difficulty is that intraspecific aggression is directed anonymously at a conspecific. Thus the inhibition of this aggression is, in higher societies, confined to personally known individuals. One can love only persons; but one can feel rage against a whole nation. The same is true to some extent in our ordinary lives. We most of us feel more rage against a culprit who does us some injury if he is unknown to us personally, then when we are already acquainted with and in some degree like him. Similarly, hunting loses its attraction to the hunter the more he gets to know individual members of the hunted species. It may be, as von Holst thought, that selection against aggressive tendencies in man is taking place fast enough to avoid world catastrophe; but I do not feel convinced that we know enough about the genetic organization of the aggressive drive in man to rely very greatly on this. If in the future a strong and deliberate selection against aggression were put into effect (and this would seem extremely hard to do as long as the world is still divided into national entities), it might change man's nature in other incalculable and perhaps undesirable ways. Another difficulty is that the basic drives of animals and men are very firmly cemented into the genetic mechanism and are extremely difficult to alter by breeding under domestication or by training. What one can do is to substitute new objectives and new ideals for old ones. This so he thought is the only reasonable hope, and I agree with him that it is reasonable. I agree with him too in the view that the ultimate power of human common sense can hardly be over-estimated. If we have that common sense expressing itself in social and natural selection then the grounds for hope are much firmer than perhaps we think. (see pp. 153-4 below),

CHAPTER 5

The Scientific Outlook and Some Moral Concepts

MORALS AND EVOLUTION

In the last chapter we were considering some of the more pressing ethical problems of today in the light of the scientific, particularly the biological, outlook. Now we must approach one or two fundamental concepts, which can justifiably be considered as moral rather than ethical, from the same point of view. Here I think I should clear the ground by saying that I certainly do not think, as apparently Sir Julian Huxley does, that we can deduce any final and precise answer to ethical problems by studying the *details* of the course of evolution. Yet it is true that the more comprehensive a viewpoint we secure of the evolutionary process, the more relevant it seems to be to such problems. Thus I referred above to the view of an eminent geneticist[1] that biological progress not only involves increasing versatility of individual species, but also increasing diversity of species harmoniously adapted to one another. I do not know of any summary of present-day views of evolution more profound than that. And it is striking how we can extrapolate it to apply with equal or greater force to human society. Progress in human society – that is to say, that which is 'good' for man – depends upon the growth and fostering of all those features which increase the diversity and richness and beauty of individual life while, *at the same time*, leading to greater unity and coherence of the whole. And it is to my mind striking that, at this level, what appears to be good for the race and community is also good for the individual. Such a prescription, of course, contains its own tensions and inconsistencies, in that two opposing tendencies have to be balanced and co-ordinated if the best is to be achieved. However, this is such a general, indeed universal, characteristic of the conditions which lead to progress that it need cause no surprise to encounter it again here.

SCIENCE AND RELIGION

But however much we may think that the course of evolution provides us with a ready made yardstick for what is ethical; there can be no temptation to consider moral concepts as such as derivable from the scientific picture of the origin and development of the living world. When we come to discuss morals we are entering the domain of theology and so it will be helpful to consider in the first place, the relation if any between the 'queen of the sciences' of yesterday and what I regard as the king of the sciences of tomorrow, namely biology. Since the cessation, in recent years, of many of the more crude and vigorous lines on which the scientific rationalist attacked religion and theology, there has been a tendency amongst innumerable writers on the subject to swing, in my view, somewhat too far to the other side. Thus it has often been said that there are no real grounds for conflict, and that the fight has all been a tragic mistake. The bases of this conclusion[2] are variously assessed. It may be said that science deals with the objective world and theology with the existential world. Or it may be argued that science discusses efficient causes and theology final causes. Or again that science is concerned with empirical truth and theology with symbolic truth. All these statements doubtless draw attention to important points which we must keep in mind; but they tend too often to attempt to smooth over and even to depict as unreal, the differences which are, or at least have been, real enough; and which it is essential should be brought into the open and thoroughly investigated. As we have seen, there are a number of discontinuities within science itself;

and these should surely warn us against the dangers of hiding discontinuities elsewhere. But it is a rather curious fact that we can find what seems the sanest and most promising view of the relation between the two, expressed in almost identical terms; on the one hand by a scientist, usually regarded as a Marxian atheist and mechanist, writing thirty-five years ago – namely J. B. S. Haldane; and on the other hand a modern theologian, well known – one might say notorious – for his disregard for, and apparent ignorance of, the scientific outlook, namely Paul Tillich. The latter says that *science is a matter of preliminary concern and theology of ultimate concern.*[3] The former summarizes what he regards as the ideal relationship of religion and science, by stating it as his view that religion is a way of life, and an attitude to the universe, which brings man into closer touch with the inner nature of reality. Although statements of fact made in the name of religion are often untrue in detail, they may often contain some truth at their core. Science is also a way of life and an attitude to the universe, but it is concerned with everything but the nature of reality. The statements of fact that it sets forth are generally right in detail, at least as a reasonable approximation; but they can only reveal the form and not the real nature of existence. Haldane[4] concludes by saying 'the wise man regulates his conduct by the theories both of religion and of science. But he regards these theories not as statements of ultimate fact but as art forms.' And it is interesting to find a writer of real power and originality and also of deep thought, namely Joyce Cary[5] arguing in effect that art itself is only worthy of attention and allegiance in so far as the artist is living in the faith that artistic values are not just the expression of feelings and attitudes, however delightful, but are in fact engendered in the process of establishing a relation with an objective reality transcending ourselves. He says (p. 156) 'when we recognize beauty in any ordered form of art, we are actually discovering new formal relations in a reality which is permanent and objective to ourselves. . . . When the impressionists devised their new technique to express and reveal their new intuition they gave us a revelation of a beauty no one had seen before. . . . They revealed a truth as a scientist reveals a new element . . . a part of the universal order of things.' And it is clearly Cary's view that an artist without that faith (which may nevertheless be subconscious only and not realized) is no true artist at all.

The differences, then, between science, theology and art are real and not to be ignored. On the other hand, links and bridges are today constantly being pointed out where not so long ago the chasms were thought to be bottomless and unbridgeable. From the scientific side, Hinshelwood has pointed out more than once how often we judge scientific theories and systems by the degree of mental and aesthetic satisfaction they afford and how, surprisingly perhaps, such criteria prove effective in leading to further scientific advance. He says,[6] 'men of science do, unconsciously at least, place hypotheses about nature in a sort of hierarchy of esteem, where the conservation of energy and the second law of thermodynamics rank high and mere working rules rank low'. So let us consider further and test the strength of whatever bridges we seem to see between the life sciences and theology – for I think that even the most dyed-in-the-wool physicist will be ready to concede that in this matter it is biology that in many ways occupies the key position.

So we come to look at some moral ideas and concepts from the biological point of view; excluding, as far as we can, preconceptions arising from other attitudes of mind. Let us be bold enough to commence with one of those all-embracing concepts of almost equal concern to a number of religions and not merely the Christian religion, namely the idea of wholeness and the related concept of perfection.

The contrasted ideas of partiality and of wholeness constitute perhaps the most ancient intellectual tools of mankind. Right through all our thinking and formulating, there runs the process of making categories and classes – categories which are exclusive or inclusive; and manipulating these in such a manner as to obtain new insights and perceive new relationships which, so we believe, often result in some new advance in understanding ourselves and the world around us. It is therefore obvious that the ideas of wholeness, of unity, and the idea that such wholes are often not primary, but are composed of the aggregation or co-ordination of lesser units, should be encountered again and again at all levels in the scientific analysis and presentation of experience. In the physical analysis of matter man has been used to this since the ideas of atomism were first put forth. In its modern form it has, of course, given rise to the physics of fundamental particles and the quantum laws. What is relatively new in the physical approach is the evidence it seems continually to be providing that the same fundamental laws are operating throughout to the farthest limits as revealed by radio astronomy. It has been argued that,[7] as more and more physical considerations are taken into account, fewer features of the universe remain accidental. Therefore we are led on to ask the question, 'Does the process ever stop?' I think there is no physicist who would be willing to give an affirmative answer to that question. Sciama at least is inclined to suspect that when everything in the *physical* universe is taken into account, there is nothing accidental left. As a physical system the large-scale behaviour of the universe depends on the laws of motion. Some[8] would add that owing to the unity of the universe, the laws of motion depend on its large-scale behaviour; and they suggest that the requirements of self-consistency may be so severe as to permit only one model of the physical universe. If they are right, and the unity and uniqueness of the universe are linked together in the most intimate way, the universe itself can be looked upon as the sole self-perpetuating consequence of its long-range interaction. By no means all physicists would of course agree with such a view of the physical world; but the very fact that it can be seriously and plausibly advanced is sufficient indication of the astonishing theoretical developments which are taking place in the cosmological field. But before we leave this subject, a word of warning is necessary: this view is the view of the physicists of the physical world – it does *not* seem to imply that the world as seen by biology is also the sole and necessary expression of the quantum laws and the laws of motion. As was argued in an earlier chapter such laws are never found to be 'broken' in biological systems; but they are not themselves relevant or adequate to 'explain' the facts of biology and psychology. If we see a similar unity there, it must have a different origin.

Turning now to the biological viewpoint, it seems to me that a similar unified picture is in the process of emerging. Not only do ideas such as heritability, sentience, homeostasis and adaptation appear to be useful throughout our studies of the living world; above and beyond this we find a unity in the fact that activities such as these and many others besides are accomplished by widely different mechanisms. In other words, closely similar machines are, in nature, constructed of widely different materials, and similar objectives are achieved by many different mechanisms.

This overall tendency towards wholeness and unity can certainly be plausibly derived from the picture presented by evolutionary biology, and most clearly is it derivable when we come to consider the higher types of behaviour in animals and the neuronal organization that goes with it. This view is a central part of the attitude of the gestalt psychologists. According to them the study of many different sections of the animal kingdom shows the integrative tendency at work whereby the whole becomes more than the sum of its parts. The very idea of gestalt as of diversity constituting unity is, as Langer[9] puts it, the keynote of rationality which though often thought of as characteristically human, lies deep in our pure animal experience. This unconscious appreciation of form is, she argues, the primitive root of all abstraction. Although much, perhaps most, of the gestalt building that goes on in the animal organism is achieved below the level which we human beings regard as conscious, it is I think now widely agreed that the primary units, with which our experience and that of the higher animals work, are not sensations but elementary perceptions – and in accepting this we are granting a major part of the position of the gestalt psychologists in so far as it concerns their ideas of wholeness and unity. From the evolutionary point of view I think we can sum up the subject by saying[10] that something like mind must be multiple in the representatives of a great many of the lower groups of animals and that the long course of evolution has, amongst other things, led to a very slow and gradual integration of all the centres of mental activity into a single one. This is the tendency which, in spite of their defects, the writings of Teilhard de Chardin have vividly and compellingly expressed; an expression which I believe has fully justified the enthusiasm with which his writings have been so widely received. The evolutionary picture shows us then that though the human mind is in many respects a feeble and fragmentary thing; this divided nature is in part at least an outcome of the evolutionary course which has led to man. However imperfect our minds may be in unitary performance, there is undoubtedly, over and above this weakness, a sense in which the mind of a man is a unity; but this unification is, evolutionarily speaking, a very late development, and even yet is by no means perfect.

With the coming of ideas and concepts appropriate to living systems, particularly with the coming of the idea of adaptation, the concept of perfection also enters science. We of course encounter the word 'perfect' in physics and mathematics; where it is used, as in the term perfect gas, to indicate a theoretical model or a close approximation to it. But in biology it means something else, or rather something more, and indicates the degree to which the process of natural selection has completed its task of securing adaptation. And if and where we seem to see trends leading in a direction of, say, greater and more complete integration, then again we bring in the idea of degrees of perfection as the evolutionary process seems to yield results which are still closer in conformation with the trends already exhibited.

Let us now consider wholeness and cognate ideas in other fields of thought and compare then with those of the scientific viewpoint. I have already quoted a leading geneticist as to what is good for a species, and have suggested that the same is good for the human individual in society. But although, as will already be apparent, I consider that much of ethics can be derived from the nature of the evolutionary process, I do not believe that to be true of the fundamental ideas of morals. There have indeed been attempts to derive them in this manner. Thus Hobhouse[11] believed that the nature of the evolutionary process was such that valid ideals will in the long run conquer invalid ideals because only valid ideals are self-consistent; and the ideals which lack self-consistency will prove 'unfit' and so suffer elimination. This is part of his view that 'the good' is the principle of organic harmony. But these conclusions as to the evolutionary process never led Hobhouse to doubt that, at the core of the religious consciousness, there are elements of genuine experience giving true insight into the real. He regards the development of religion as the progressive apprehension of the spiritual. But some modern thinkers find that reflections on the nature of the evolutionary process carry them to a very different conclusion. Thus Huxley[12] suggested that the basis for the quality of absoluteness and otherworldliness possessed by the super-ego is to be found in the all-or-nothing methods, leading to severe repression, adopted by animals in conflict – although he draws back from following this to its logical conclusion when he speaks of 'the moral nobility of personality, a sense of oneness with something beyond and larger than ourselves, which is either moral or transcends morality'. Waddington[13] derives morality from the nature of human society which depends upon a socio-genetic system for transmitting and accepting information from one generation to the next, so that we are by nature acceptors of authority. He sees the belief in other-worldliness and absoluteness, our demand for certainty in beliefs, as largely the outcome of the break-up of the harmonious state of solipsism characteristic of early babyhood. And he seems to regard this as the complete explanation of all our beliefs in, and craving for, ultimate values. For myself, I can agree that natural selection, acting on 'human' societies, might lead to situations in which more pervasive sympathy and stronger altruism might prevail.[14] But there does not seem to be any clear reason why this should have gone beyond the requirements of the social conditions favourable to the development of ethics. But all concepts involving absolute value must, I think, be treated as scientifically unverifiable. And this, I take it, was essentially the view of Hobhouse in this matter.

In theology, the mystical awareness of God as the whole, in some sense the unity of all things, is in effect the basis of all the fundamentals of belief. And it is this awareness from which derives all the power and compulsion of religious assent to the belief that God is love. For love is the only human quality through which we can begin to apprehend the implications of the divine in all their vastness.

It is for this reason, because I believe they are emphasizing the most profound and compelling insights into the thought of man and in the Christian revelation, that I value Bishop Robinson's[15] popularization of the views of Tillich – not to mention other more recent theologians. It seems to me that he is speaking to the needs of the newly awakened and inquiring intelligence which is a feature of our present day society. This it is, I am sure, the duty of the Christian churches to perform; a duty which too often in the past they have failed to do, or have done inefficiently. I regard it as one of the most encouraging signs of religious life today that it is now being done, and being done in this manner. But *Honest to God*, valuable though it is, is only a tentative 'first word' which omits much that is essential and distorts much else. It will be a long time before the 'last word', written so that the 'plain man' can understand, is even approached.

In this view of love as the individual expression of the unity of all things in God, there is something very close to the scientific ideal of integrity; of devotion to the truth wherever it may lead. It is indeed heartening to find modern theologians emphasizing this in terms which could be equally agreed by existential philosophers and by scientists. Thus Woods[16] shows how existentialists make a distinction between a way of living which is

138

authentic and one which is un-authentic; between a way of living in good faith or in bad faith. He goes on to say 'the basic misconduct is to live without sincerity and integrity'. This is the best definition of sin I know; and perhaps it is the only valid definition of sin. For sin is essentially that which opposes the urge towards wholeness and unity, which is the love of God; and denial of this leads not to integration but to disintegration.[17] Here there is a very profound biological parallel; it can be argued, I think with full justification, that in psychology stimuli and experiences which are painful, if they are too prolonged and intense, result in the breakdown of integrative mechanisms, considered both neurologically and psychologically. This is what evil is, from the biological point of view. As so often, Dante here seems to come nearest to the perfect expression where he says[18] 'but human kind is most like God when it is the most one; for the true principle of unity is in Him alone'. And again[19] Dante suggests that even the saints may have but '... feeble sight unable to detect the first cause *whole!*'

But the injunction to love the highest, while fundamental, is a bleak, bare and in its way partial, statement of what the Christian religion is about. Adhering to this alone can easily lead, in those who are but little aware of the hidden motives and complexities of their own nature, to a self-centred and priggish care for one's own soul, a belief in the desirability of 'nurturing one's spiritual resources' (to quote a horrible phrase recently much in use in certain religious quarters), which is stultifying to one's spiritual growth and alienating to one's fellows. As Williams[20] says the process of growing up into self awareness, recognizing our fears and guilt over what is buried within us for what it is, results in a maturity from which flow all the highest achievements of human life; of which the greatest is the capacity to give oneself away in love. And self-effacing love, risking the loss of our all, so it sometimes seems of our very soul, is the very core of the teaching of Christ. All this presents the specifically Christian view. But it is, I submit, this awareness, often subconscious, which is the power behind the scientific conscience and scientific unselfishness. It can appear as saintly as any other kind.

SCIENCE AS A RELIGIOUS ACTIVITY

Woods[21] claims that natural science can never *of itself* provide adequate grounds for moral decisions. My contribution here to this view is my attempt to demonstrate that the task of practising science shows increasingly that the enterprise can only proceed by consciously or subconsciously adopting an attitude to the sum of things which is essentially a religious attitude; essentially – even at a very long remove – a search for the unified reality underlying all experience. Nor is this attitude confined to theologians and scientists: it is characteristic also of much if not all that is best and most profound in both philosophy and literature. Thus Ludwig Wittgenstein[22] expressed his conviction that one has duties from which one cannot be released, even by death. And Dostoevski[23] said, 'If God did not exist, all would be permitted.' I think Baillie is correct when he sees in this kind of statement whether coming from scientists, philosophers or artists, the ultimate ground for the sense, which all share, of the reality of the corporeal world, proofs for which are so unsatisfactory. This sense is dependent upon the apprehension of the world as shared, in some sort derivative from the sense of reality of other selves and above all of a sustaining spirit.

Much of what I have just been saying, though certainly not all, is applicable to deism as well as theism; and indeed to many non-Christian religions. For the deists everything without exception was regulated by natural law. But I do not believe it is possible in this day and age to find in the scientific venture, and in the dedication of scientists to it, anything which remotely supports deism against theism. In this matter I cannot do better than summarize in a few words the recent conclusions of Professor Ninian Smart.[24] First, if theism be true, it brings with it a belief in creation. True, the deist may also have a belief in creation but certainly not, I think, in creation by emergent evolution and still less a belief in nature as sacramental and as an avenue of religious awareness. So the Christian view confronts both the uncompromising monotheism of Islam which finds the concept of incarnation altogether blasphemous; and on the other side it faces the multiplicity of incarnations characteristic of Hindu belief. Secondly, it is true that cosmology certainly does not rule out the possibility of recurring cycles of creation, a traditional eastern view, and cosmology certainly agrees with other strains of oriental thought which considers the universe as of infinite duration of time. But as Smart points out, the eastern view of the cyclical nature of history is partly dependent on belief in rebirth. And in so far as modern genetical ideas have anything to say upon the subject, they are for mathematical reasons inalterably averse to the concept of reincarnation – the individual's gene complex is unique, never to be precisely repeated in the future of the world. Thirdly, in many eastern countries there is now a tendency to a reinterpretation of human history in a manner more consonant with the Christian vein of thought, though not of course openly acknowledging it. It arises out of the fact that after centuries of social stagnation there is a drive towards political and economic freedom which is beginning to touch the awareness of everyone. Fourthly, evolutionary theory, and the ideas upon the significance and implications of evolution which I have been discussing, provide perhaps the strongest scientific evidence of all in favour of a theistic over a deistic interpretation. It reinforces the sense of a direction, and I would say of progress, though here perhaps Professor Smart would part company with me.

Any further discussion of the relation between biology and those fundamental ideas which lie at the basis of moral and ethical judgements must, I think, be largely based upon the idea of natural religion, and the consideration of its relation to 'revealed' religion, in particular to Christian theology. In considering this subject, it is necessary to realize that in the last hundred years there has been a change in the generally accepted meaning of the term natural religion. Cardinal Newman,[25] writing in 1870, regarded the conclusion as obvious that Christianity is merely the completion and supplement of natural religion and of previous revelations; and he shows that this was what Jesus and his apostles taught. But natural religion for Newman implied knowledge coming through our own minds, through the voice of mankind and thirdly through the course of human life and human affairs. He was not, I think, in any way concerned with the view that science as an objective study of natural phenomena, let alone the theory of evolution by natural selection, could provide any unique or necessary knowledge of the ways of the Creator. Still less did he imagine that the objective study of mental processes and of human and animal behaviour, which constitutes the discipline of psychology today, could lead to knowledge at least as important and profound as that obtained through introspection.

Towards the end of the nineteenth century and in the first decades of the present century, there was a strong and encouraging tendency in the direction of acceptance, by theologians and philosophers, of the implications of evolution theory for religion. This was exemplified particularly by the writings of Whitehead, Lloyd Morgan, Pringle Pattison, Alexander and Stout, amongst philosophers, and Gore, Barnes, Tennant, Oman and Raven amongst theologians. But it was one of the disasters of the First World War and of the forty years that followed that this far-seeing and enlightened tendency amongst theologians was largely smothered. I believe Canon Raven [26] is right when he speaks of the great blight which descended upon theology when it accepted the 'religion of despair' of Karl Barth and Reinhold Niebuhr; and I think he hardly exaggerates the disastrous effect of this when he speaks of Christ's religion becoming 'almost wholly confined to scripture treated as the single means of revelation and to liturgy as the single activity of the church'. I therefore rejoice to find a distinguished theologian saying very recently[27] 'intellectual objections to Christianity nowadays, and the fact that there are at present no convincing answers to them, in my judgement, both grow out of one root. This is that there is no general or widely accepted natural theology. I know that many theologians rejoice that it is so, and seem to think that it leaves them free to recommend Christianity as divine revelation. *They know not what they do*, for if the immeasurably vast and mysterious creation reveals nothing of its originator, or of his or its attributes and nature, there is no *ground* for supposing that any events recorded in an ancient and partly mythopoeic literature and deductions from it can do so. . . . The only possible basis for a reasonably grounded natural theology is what we call scientific.'

If then we accept Newman's view of Christianity as the completion and supplement of natural religion, but giving the latter term its modern connotation, to what extent can we argue that specifically Christian moral concepts are a fulfilment of ideas generated by natural science and receive support from it? I think here the first concept we must consider is one that is a notable hindrance to not a few deep-thinking biologists, namely the idea of God as a person. Now it is true[28] that no thinking man can ignore the fact that the universe *as modern astronomy reveals it* gives no indication of personal activity. Men tend to think that the originator of inconceivably vast masses of flaming gas cannot be personal in any ordinary meaning of the word. But surely there is an obvious error here; an error into which the ordinary man is very prone to fall, and which is shown up by the words '*as modern astronomy reveals it*'. It is of course true that the universe is only *very partially* revealed by modern astronomy. It is the old and very human error of making obeisance to mere size. If on the contrary we look at the universe from the point of view of the life sciences, including psychology and sociology, it is utterly impossible to say that it reveals no sign of personal activity. But of course a much more serious difficulty underlies this very general reaction I have been speaking about: this is that the very concept of personality as implied in ordinary speech, involves limitation. And this is true not only of everyday speech. Baillie[29] points out that much modern philosophy had taken for granted that personality is a characteristic mode of being of finite spirits, the very hall mark of our finitude. It had therefore been very naturally contemptuous of all theological attempts to extend such a characteristic by way of analogy to the divine mode of being. But in fact more than a hundred years ago this kind of assumption was powerfully controverted by the philosopher Lotze and the philologist Ritschl who argued that personality was an ideal conception never more than very imperfectly realized in human existence. The modern acceptance of this view has received what I think can be regarded as definitive expression by Webb.[30] This can perhaps be baldly summarized as the doctrine that there is personality in God. But some modern theologians, e.g. Montefiore,[31] find they cannot go even as far as that and must content themselves with saying that God works in a personal way. I myself find this statement inadequate, preferring the way in which Webb formulates it; but I particularly like Besant's[32] version of this when he says 'with one accord theologians acknowledge that we only speak of God as personal because that is the highest category of unity in diversity which we know'.

There is another aspect of this matter which seems to me of great importance. Woods[33] has recently given a new expression of it. He argues that there is an irreducible distinction between seeing that something is so and seeing that it ought to be so. The two judgements cannot possibly be the same. There is a fundamental Christian conviction that somehow there is a realm in which what ought to be is what is; and I believe he is right when he argues that it is impossible to conceive moral judgements being made where there is no moral agent; and a moral agent implies personality.

Now it is obvious that in all discussions of this kind we are constantly frustrated by the inadequacies of language. We are continually being driven towards an awareness of the absolute or the transcendent, and continually becoming tongue-tied in our attempts to express it. Woods in his discussion of the transcendent argues that besides having experience of the temporal and the

transient, we have at least as strong a sense of the unchanging be-
neath the changes which we see. This amounts to a vision of
nature or the natural order which survives throughout all natural
changes.[34] Woods points out that the vast majority of those who
have lost, or who have never had, any effective belief in God, may
continue to believe in nature in this sense, and this I think includes
nearly all scientists. In a previous chapter I discussed the idea of
communication and what it is we communicate, and here where
we come up against the impasse provided by the idea of the trans-
cendental and our inability to communicate it, we return to
the subject again. It is after all no wonder that language begins to
fail us. Dante deals with this impasse in *Il Paradiso*[35] and tells us,
what we can well understand from a perusal of the other mystics,
how it is not merely language which failed him in experiencing
the ineffable, but even memory itself. That language should fail us
need in fact cause the student of evolution and biology no surprise; for all ideas that represent any new advance of human
thinking come to us first in wordless form – in advance of language, so to speak. Only later do they become expressible and
then perhaps only in part. We see this clearly in the process of
evolution of animal communication and it is surely inconceivable
that the same process was not occurring in the course of the origin
and evolution of human language. To argue that the proto-
hominids lacked all concepts such as 'push', 'pull', 'help', 'danger',
'come', 'go' and even 'good' and 'bad' until they got words for
them seems so preposterous a view as not to require further consideration. It is a serious criticism of many linguistic philosophers
that they appear to lack almost completely the evolutionary outlook.

TRANSCENDENCE

Any discussion of transcendence and the failure of methods of
communication leads naturally to a consideration of mysticism
and its relation to the scientific picture. First I think we should

resist strenuously the view which regards mysticism as an ecstatic and probably pathological state experienced by a very few very unusual people. Still more should we be chary of dismissing it as essentially something spurious; the product of charlatanism. On the contrary, the term should refer to the awareness of values in part at least above and beyond the scope of current symbolism to express – in particular the intense awareness of the whole as the unity of all things.[36] I consider, in fact, that a scientific and dispassionate consideration of the phenomena demolishes almost completely the argument of those critics who have attributed mystical raptures to autosuggestion, to compensation for thwarted ambition or to aberrations of sex passion. On the contrary, I believe that the mystic differs in degree but not in kind from more ordinary men. All recognition of value, all appreciation of beauty, truth and goodness testifies to the emergence in us of the eternal – though in our consciousness of such emergence, as in our response to it, we attain various levels of completeness.[37] Dean Inge[38] regarded mysticism as only a further development of a universal religious practice, that of prayer. The characteristics of the true mystic include perspective and proportion, sanity and stability. The test of a true mystical ecstasy lies not in its outward sign, but in its inward grace;[39] and this distinguishes it sharply from the effect of any external or purely physiological technique for enhancing the perceptions. In the true mystic, external circumstances and of course ascetic techniques which may, indeed must, have certain physiological secondary effects, may well enhance its vision but will not alter its fundamental nature. Professor Zaehner[40] has I think disposed finally of the kind of superficial and spurious argument provided by Mr Aldous Huxley in his *Doors of Perception*. But it is not only the religious and the 'sensitives' who witness to the reality and profundity of the mystic illumination. Thus Professor Broad[41] while claiming himself to be almost wholly devoid of mystical or religious experiences, combines this with a great interest in such experiences, and a belief in their extreme importance in any theoretical interpretation

146

of the world. He says 'probably most people have them to some extent . . . it seems reasonable to suppose that the whole mass of mystical and religious experience brings us into contact with an aspect of reality which is not revealed in ordinary sense perception, and that any system of speculative philosophy which ignores it will be extremely one-sided'. And it is in line with this assessment of Broad that we find that not only many scientists but philosophers as diverse as Bergson, Bradley, Mctaggart, Whitehead and Wittgenstein have found that the most strenuous attempts to carry logical thought to its utmost limit culminate in mysticism. Then again artists who ponder the foundations on which their work rests tend[42] to accept the views of Benedetto Croce, that the basis of art is a type of intuition which at once makes it a universal activity by which it can achieve direct knowledge of the world.

SCIENCE, ART AND RELIGION AS WAYS OF KNOWING

The view that I am putting forward, that science, art and religion have something in common as a way of knowing, is essentially that which Professor Polanyi,[43] sets forth – stating it as his view that human knowing can only be achieved by the commitment of a personal will and judgement. In its highest expression this is the faith which takes the form of an intellectual passion to comprehend and understand, which cannot be gainsaid and which is perhaps the strongest driving force in the intellectual activity of man. For Polanyi, the development of this way of knowing is not a purely human faculty but is the culmination of a trend clearly evident in the course of evolution. This brings him in some respects closely in line with Teilhard de Chardin (though as a savant greater than Teilhard) in that he sees these various ways of knowing, which at heart are all part of one way, as continuing and subject to further evolutionary development, the limits to which no man can guess.

CONVERGENT INTEGRATION

Teilhard's vision of the possible future of evolution involves the development of mankind into a single psycho-social unit, implying that we should consider inter-thinking humanity as a new type of organism, whose destiny it is to realize new possibilities for evolving life on this planet; a form of convergence or 'complexification' resulting from the fact that in modern scientific man evolution is at last becoming conscious of itself. This process we can designate by Julian Huxley's translation as '*convergent integration*'. The idea of evolution continuing is not, of course, just a phantasm of a woolly-minded mystic but the considered view of some of the most eminent geneticists of the time.

To suppose that evolution, whether of animals, plants or men, has now come to an end is unwarrantable. Such a belief flies in the face of innumerable facts. First, in regard to selection of the orthodox kind, the raw material for selection is still provided by the human organism just as it has always been. Events of recent years force us to dwell very deeply and persistently on those drastic mutations, which we see too often, which produce in human beings fatal hereditary diseases and spectacular malformations. But Müller[44] supposes with good reason that small mutations are about five times as frequent as drastic ones. The average mutation rate per gene in human beings is probably more than 10^{-5} rather than less. To take the extremely conservative view, let us suppose[44] that the two sex cells with which the life of every human being begins contain only 20,000 or 2×10^4 genes. It will then be a conservative estimate that $2 \times 10^4 \times 10^{-5}$ or close to 20 per cent of all people will carry one or more newly arisen mutant genes. Mutants, then, are evidently not at all rare in human populations, nor are they rare in other organisms. The basic condition of evolution by natural selection is still with us. But having said this, we must again make it clear that human evolution is no longer a matter of the simple conservation and retention of new variants by a crude natural selection. Human evolution is now

148

biology plus culture, and any future view of the human race which forgets this is bound to be wide of the mark.

And indeed cultural evolution has largely taken over already. All healthy individuals[44] of *Homo sapiens* have a capacity to learn a language, any language, and to acquire a culture – any of the cultures of any group of people anywhere. This can be said to be one of the biological characteristics of our species – as characteristic as the erect posture and our non-opposable big toe. There are certainly no genes for the English, the Chinese or the Bushman language or culture. Our plasticity is in fact a species-trait, formed by natural selection in biological evolution. But as the same authority points out, it is a fallacy to think that specific traits do not vary or are not subjected to genetic modification. Genotype plasticity does not preclude genetic variety. There may be variations in the degree of plasticity; some of the functions or roles which exist within a culture may be more congenial and hence more easily learned than others. And it certainly cannot be proved beyond doubt that there has been any appreciable change in human mental ability during the last fifty thousand years. As Dobzhansky says, we cannot plant some identical twins to be reared by Peking man or by the Neanderthalians and leave co-twins to grow up in a modern society. Emergence and development of culture makes adaptation to changing environments by means of genetic changes less binding than it was in animals in a pre-cultural state. Man did not need to grow warm fur to cope with cold climates simply because he donned warm fur garments. But in the end there is no way for culture to ward off genetic change altogether. Culture does not make human environments stable and uniform. Given environmental flux, the necessary and sufficient condition for genetic change is the availability in populations of genetic variants, some of which are better and others less well adapted to shifting environments. Selection will do the rest. It will multiply the favoured variants and depress or eliminate the unfavourable ones.

There certainly seems to be a strong genetical argument for a

F

development of society which tends *slowly* to break down pre-existing castes and stratifications. Equality decreases the wastage of the genetic potential of any species, whether human or animal. In man it favours manifestation of talents which remain hidden in societies that let high cultures and refinements flourish while a great majority of people live in misery and ignorance. The same authority suggests that although this cannot be proven rigorously, it is extremely probable that men like Leonardo, Beethoven, Darwin or Einstein did possess what he calls rare and precious constellations of genes. If these gene constellations had appeared amongst Indian untouchables or Negro slaves, or even in the slums of big cities, their carriers might not have accomplished much, and humanity would have been the losers. Equality of opportunity is an ideal not uniformly appealing to everyone, and many have expressed fears that as class competition decreases, competition between individuals increases in intensity and vindictiveness. It has been suggested[45] that the complete elimination of classes would mean the installation of a dog-eat-dog society; and that mankind's biological as well as his cultural welfare demands competition of many separate class or race populations. Some of these will become extinct, others will survive and repopulate places left vacant by those who have succumbed. I agree with the criticism that the basic error of these views lies in the implied assumption that equality of opportunity and mitigation or elimination of inter-group competition inevitably result in uniformity, in levelling and in the disappearance of genetic and cultural variety. But this is not necessarily the case at all. The variety of human genotypes and hence of inclinations and capabilities is increased not decreased by hybridization, and the same may be true on the cultural level also (see p. 118 above).

It would of course be foolish to suppose that even the most expert geneticists, the most accomplished students of human social organizations, can see more than a very minute distance into the future. Evolution shows us abounding and surely unpredictable developments, cosmic processes overcoming their own bounds in producing life, biological evolution transcending itself in giving rise to man. It is only common sense to suppose that human evolution may yet ascend to a superhuman level, and that the creation is not complete. It is not an event which happened in the remote past, but rather a living reality of the present. Creation is a process of evolution, of which man is not merely a witness but a participant and partner as well. In putting forward such views modern geneticists are in fact saying what Teilhard de Chardin, as a palaeontologist, was saying quite independently. How limited, how indifferent and how ill-informed (even for 1935) was Bertrand Russell who said 'there is no law of cosmic process but only an oscillation upward and downward with a slow trend downward on the balance owing to the diffusion of energy. This at least is what science at present regards as most probable, and in our disillusioned generation it is easy to believe. From evolution so far as our present knowledge shows, no ultimately optimistic philosophy can be validly inferred.'[46] It is gratifying indeed to be able to say that the modern biological picture in no way justifies such a Jeremiad. For biological prediction we do better to go to the biologist and not to the physicist or mathematician.

In response to some misguided critics of Teilhard, it is pleasant to be able to counter them with a quotation from J. B. S. Haldane who, in an article published thirty years before,[47] says: 'Now if the co-operation of some thousands of millions of cells in our brain can produce our consciousness, the idea becomes vastly more plausible that the co-operation of humanity, or some sections of it, may determine what Comte called the Great Being. Just as, according to the teachings of physiology, the unity of the body is not due to a soul superadded to the life of the cells, so the superhuman, if it existed, would be nothing external to man, or even existing apart from human co-operation. But to my mind the teaching of science is very emphatic that such a Great Being may be a fact as real as the individual human consciousness, although, of course, there is no positive scientific evidence for the existence of such a being. And it seems to me that everywhere ethical experience testifies to a super-individual reality of some kind. The good life, if not necessarily self-denial, is always self-transcendence. This idea is, of course, immanent in the highest religions, but the objects of religious worship retain the characteristics of nature-gods or deified human individuals. It was more satisfactorily expressed by Comte; but there is much in Positivism as originally conceived by him which seems unnecessarily arbitrary.'

This concept of revelation to come is of course held in the doctrine of the Holy Spirit.[48] But as often, it seems needful to come back to the scientist for a wider more inspiring statement. Here is Polanyi again: 'Christianity is a progressive enterprise. Our vastly enlarged perspectives of knowledge should open up fresh vistas of religious faith. The Bible, and the Pauline doctrine in particular, may be still pregnant with unsuspected lessons; and the greater precision and more conscious flexibility of modern thought, shown by the new physics and the logico-philosophical movements of our age, may presently engender spiritual conceptual reforms which will renew and clarify, on the grounds of modern extra-religious experience, man's relation to God. An era of great religious discoveries may lie before us.'[49] And in this connection it is most significant that, as Raven[50] shows, Teilhard in his whole vision of the future of man is actually and avowedly restating the theology of St Paul as this came to its fullest expression. Both Polanyi and Teilhard understand that Paul in the last three epistles has a vision which neither he himself, nor perhaps anyone prior to our own day, has been in a position to comprehend as fully as can now be done.

But it is only too true that, concurrently with becoming conscious of and capable of controlling his own evolution, man has at the same time attained the power of preventing it – of destroying himself and perhaps indeed the whole living world. There are today overwhelming arguments for believing that general nuclear war would be the greatest evil which could possibly overtake our race. Perhaps since John XXIII spoke so incisively, there will at last be an end to the blasphemous pronouncements of Anglican and Roman bishops and American divines, that it may be God's will that man should by his sin destroy both himself and God's creation. To my mind, because I have faith in the possibility of vast social advances yet to come and of continuing evolution as an expression of God's will for and guidance of the world, I regard the suggestion that it may be the Christian duty of man, under any circumstances whatever, to initiate or acquiesce in the murder of mankind and the living world, as the most obscene blasphemy that the mind of man can conceive. That surely is the supreme apostasy for modern scientific man. Many know not what they do; but some must know full well what they do. We are at the threshold of the human story; yet some are prepared to bring it now to a close. Wilfred Owen points to pride as being the greatest of all sins.

'Near Golgotha strolls many a priest
And in their faces there is pride
That they were flesh-marked by the Beast
By whom the gentle Christ's denied

'Lo an angel called out of heaven to Abram
. . . Lay not thy hand upon the lad . . .
Offer the ram of pride instead of him.
But the old man would not so, but slew his son –
And half the youth of Europe, one by one . . .'

The centre-piece of Teilhard's view of evolution [51] is that man, as a reflective being, is different from all other animals. This means that his future psychological and social evolution can only proceed by 'infolding'. [52] Instead of fanning-out into new environments and to form new lines, which is the characteristic pattern of animal evolution, and of human racial differentiation so far, the only and 'inevitable' course open to the races and cultures of man in the future is to become more and more unified: To come to express in fact the supreme ideal of diversity within unity. 'Converging branches do not survive by eliminating each other; they have to unite.' 'Everything that formerly made for war now makes for peace.' This, the only 'direction' which man can now take, the direction which he *must* take or perish from the earth, is ultimately made easier and more urgent by the two supreme problems of our day – the population explosion and the role of nuclear energy. The only possibility for man, ever more tightly packed on this planet is 'involution'. This co-operative process could and should be immensely aided by modern methods of transport and communication (and perhaps also by the potentialities of the computer revolution). The atomic discoveries of the past 25 years are also forcing the pace and, if we can but see it, in the same direction.

'For all their military trappings, the recent explosions . . . herald the birth into the world of a Mankind both inwardly and outwardly pacified.' But Teilhard goes on to say that the great enemy of the modern world is *boredom*. '. . . despite all appearances mankind is bored. Perhaps this is the underlying cause of all our troubles. We no longer know what to do with ourselves. Hence in social terms, the disorderly turmoil of individuals pursuing conflicting and egoistical aims; and on the national scale, the chaos of armed conflict in which, for want of a better object, the excess of accumulated energy is destructively released. . . .' But he considers that, thanks to atomic weapons, 'it is war, not mankind, that is destined at last to be eliminated.'

To us in 1965 this may sound the wildest optimism; and indeed the outcome of past policies based on the view that stronger armaments will prevent war, gives cause enough to doubt. But it must at least be remembered that history is not repeating itself in the present situation. There has never before been anything really comparable with the Hydrogen bomb and, therefore, nothing strictly comparable with our present plight. But Teilhard's reasons for his faith are arresting. He says '. . . war will be eliminated at its source in our hearts because, compared to the vast fields for conquest which science has disclosed to us, its triumphs will soon appear trivial and outmoded. . . . The atomic age is not the age of destruction but of union in research.'

But Teilhard goes on to say that dreams of millennary felicity are not what he means by 'peace'. 'Although, as I believe, concord must of necessity prevail upon the earth it can by our premises only take the form of some sort of tense cohesion, pervaded and inspired with the same energies, now become harmonious, which were previously wasted in bloodshed.' 'In short the final effect of the light cast by the atomic fire into the spiritual depths of the earth is to illumine within them the over-riding question of the ultimate end of Evolution – that is to say the problem of God.'

Wilfrid Owen writing in the Flanders of 1917 said 'All the poet can do to-day is warn'. But now, if with this great hope and faith before us, the poet, the scientist, the religious and all men of good will work together we can do more than warn. We can, as Teilhard says, animate the '. . . soul of Mankind resolved at all costs to achieve in its total integrity, the uttermost fulfilment of its powers and destiny'.

Libera me, Domine, de morte aeterna, in die illa tremenda

Notes

CHAPTER I

[1] Iwanowski, D. (1892), Beijerinck, M. W. (1898): Tobacco mosaic virus. Loeffler, F., and Frosch, P. (1898) Foot and mouth disease. For full references and discussion see Stanley, W. M. and Lauffer, M. A. Ch. 2 of *Viral and Rickettsial Infections of Man,* ed. Rivers, T. M., London (1948).

[2] Waddington, C. H. (1961), *The Nature of Life,* (see pp. 26 and 54).

[3] Picken, L. E. R., summarized in Thorpe, W. H. (1961), *Biology, Psychology and Belief,* 14th Stanley Eddington Memorial Lecture, Cambridge.

[4] Keilin, D. (1959), Leeuwenhoek Lecture *Proc. Roy. Soc. B.,* **150**, pp. 149–91.

[5] An open system is one which continually gives up matter to the outer world and takes in matter from it, but which maintains itself in this continuous exchange in a *steady state*, or approaches such steady state in its variations in time.

[6] For a discussion of recent developments in biochemistry, many of them associated with the work of Szent-Gyorgyi, see Kasha, M. and Pullman, B. (1962), *Horizons in Biochemistry,* New York.

[7] Picken, L. E. R. (1955), *School Science Review,* **37**.

[8] Schrödinger, E. (1944), *What is life?* Cambridge.

[9] Wigglesworth, V. B. (1945), '*Growth and Form in an Insect*', in *Essays on Growth and Form,* Oxford.

[10] Gillespie, W. H. (1859), *The Theology of Geologists as exemplified in the cases of the late Hugh Miller and others,* Edinburgh.

[11] Gosse, P. H. (1857), *Omphalos: an Attempt to Untie the Geological Knot,* London.

[12] Russell, B. (1935), *Religion and Science,* London.

[13] Waddington, C. H. (1961), op. cit., p. 77.

[14] Fisher, R. A. (1950), *Creative Aspects of Natural Law,* A. S. Eddington Memorial Lecture IV: Cambridge.

[15] There seems to be wide agreement amongst modern physicists who write on philosophical aspects of their subject, that causality, at least in the form that would provide for a completely deterministic view of the whole universe, is no longer tenable. Thus models of particles in motion are no longer adequate as a universal explanatory principle either in physics or biology. (See Schrödinger, *Science and Humanism: Physics in our time* (1951); R. A. Fisher (1950), *Creative Aspects of Natural Law;* E. H. Hutten (1961) 'Physics and biology', *Brit. J. Philos. Sci.,* **11**, pp. 101–8; K. R. Popper (1950) on the question of indeterminism in quantum and classical physics, *Brit. J. Philos. Sci.* **1**, 117–33, 173–95). For an acute discussion of the nature of causality in contemporary physics see Schlick, M. (1961), *Brit. J. Philos. Sci.,* **12**, pp. 177–93, 281–98. Causality in contemporary physics. Elsasser (1958) (discussed below) overcomes this difficulty by introducing a new kind of causality at the biological level.

[16] Waddington, C. H. (1961), op. cit. p. 19.

[17] Waddington, C. H. (1961), op. cit.

[18] Pantin, C. F. A. (1951), *Organic Design. Adv. of Sci., London,* **8**, pp. 138–50.

[19] Pantin, C. F. A. (1951), *Organic Design. Adv. of Sci., London,* **8**, pp. 138–50.

[20] Dobzhansky, T. (1962), *Mankind Evolving: The Evolution of the Human Species,* London.

[21] Waddington, C. H. (1961), op. cit.

[22] Waddington, C. H. (1961), op. cit. pp. 89 and 91.

[23] Dobzhansky, T. and Ashley Montagu, M. F. (1947), Natural Selection and the Mental Capacity of Mankind. *Science,* **105**, pp. 587–90.

[24] Trotter, W. (1919), *Instincts of the Herd in Peace and War,* London.

[25] This is implied in a modern version of the argument (due to Thoday, 1958) in which it is described as the power to increase individual versatility without decreasing genetic versatility.

[26] Polanyi, M. (1966), Terry Lectures Yale University (in MSS) Ch. 2, *Emergence,* p. 11.

[27] However, if we admit that the physical materials are 'designed' for

living mechanisms, in the sense of being 'fit' for the construction, then we are forced to say that, in so far as this is true, the 'purpose' of the physical universe is revealed in the living universe, i.e. that the physical unit is designed so that at some stage, somewhere, life will be possible.

[28] Bertalanffy, Ludwig von (1952), *Problems of Life: an evaluation of modern biological thought,* London.

[29] Bertalanffy, L. von (1952), op. cit.

[30] Waddington (1961), op. cit.

[31] Whyte, L. L. (1965), *Internal Factors in Evolution.* London.

[32] Harris, E. E. (1965), *The Foundations of Metaphysics in Science.* London.

[33] Quastler, H. (1964), *The Emergence of Biological Organization.* New Haven and London.

CHAPTER 2

[1] Polanyi, M. (1962), Terry Lectures, Lecture 2, p. 15.

[2] Pantin, C. F. A. (in press) Tarner Lectures.

[3] For a valuable recent summary of this see Frisch, O. R. (1964), *Causality. The Listener*, **72**, pp. 83–4. He says 'Many people say that it is hard to imagine events happening without cause, purely by chance. I sympathize with that feeling; but I do not think that it can be backed up by any real arguments.'

[4] Crick, F. H. C. (1958), The biological replication of macromolecules. *Soc. Exp. Biol. Symp.* No. **12**, pp. 160–1.

[5] Picken, L. E. R. (1961), in *The Cell and the Organism*, ed. Ramsay, J. A. and Wigglesworth, V. B., p. 95.

[6] A fuller statement of the arguments of the next few paragraphs will be found in Thorpe, W. H. (1963), Ethology and the coding problem in germ cell and brain. *Z. Tierpsychol*, **20**, pp. 529–51.

[7] Setlow, R. B. and Pollard, E. C. (1962), *Molecular Biophysics,* London.

[8] Setlow and Pollard (1962), op. cit., p. 66.

[9] Setlow and Pollard (1962), op. cit.

[10] Setlow and Pollard (1962), op. cit.

[11] Dancoff, S. M. and Quastler, H. (1955), 'The Information content and Error Rate of Living Things' in *Information Theory in Biology* (Quastler H. ed.), Illinois.

[12] Curtis, A. S. G. (1963), The cell cortex. *Endeavour,* **22**, pp. 134–37.

[13] Elsasser, W. M. (1958), *The Physical Foundation of Biology: an analytical study,* London and New York.

[14] Hydén, H. (1962), 'A molecular basis of neurone-glia interaction,' in *Macromolecular Specificity and Biological Memory,* ed. Schmitt, F. O., Cambridge, Mass. Professor Hydén (pers. comm.) tells me that his estimate is based on an assumed input – complicated or simple – to be learned, of one bit every 0.1 sec. for thirty-five years.

[15] Dixon, M. and Webb, E. C. (1958), *Enzymes.* London.

[16] Mills, W. H. (1932), 'Some Aspects of Stereochemistry'. Presidential address, Section B. *Report Brit. Asscn. Adv. Sci.,* pp. 37–56.

[17] Dixon, M. and Webb, E. C. (1958), *op. cit.*

[18] It is sometimes argued that this is a false question and that there is no sense in speaking of information content except in relation to a communication channel of some kind. Thus once a system has been ordered, the attempt to describe that order by information theory is inadmissible: information only makes sense when instructions are being conveyed from one system to another. That this criticism is in fact invalid I think is apparent from the work of L. Brillouin (1962), *Science and Information Theory*, 2nd ed. Acad. Press). Here he, like many other writers, uses the term to cover the degree of organization of a static system. It is of course true that the amount of information required to specify a system will depend on the scale or size of grid used. In the ordinary case of transmission of information, this scale is given by the conditions of the communication channels and the speed with which they can work. But when one speaks of the information content of a system without specifying a grid or channel then one is doing so in relation to a fully randomized state and one is then using one's estimate as an estimate, however approximate, of the degree or order or disorder of the system. Now in such a situation estimates of information are bound to be far less definite and more questionable than if one has a fully specified channel on which to base one's calculations. If one has the channel specified and the nature of the units of which the system is constructed also specified, then one can estimate relatively accurately the amount of information required to put into the system the necessary further degree of order. This is to say that in the normal case we are making some estimate of the order already present and calculating the degree of order to be further added to give an observed result. When one cannot fully specify a channel then one's estimate of information content is likely to be far less accurate because one's assumptions as to the initial state of disorder are themselves far less soundly based. Nevertheless, the information content of the system is a meaningful expression and is necessarily introduced in discussions of the origins of order from disorder.

A further difficulty may be alluded to here. The use of radioactive tracers has confirmed the view that all the amino acids of the protein molecules in the living organisms are exchanged sooner or later. As Elsasser (op. cit.) says, 'It is doubtful whether in the course of several months there remain more than a few amino acids in a biologically active tissue that have not changed their position, structure, or chemical environment.'

[19] Quastler, H. (1959), *Quart. Rev. Biol.*, **34**, p. 228.

[20] Raven, C. (1961), *Oogenesis*, Oxford.

[21] Commoner, B. (1962), Is DNA a self-duplicating molecule? In *Horizons in Biochemistry*, ed. Kasha, M. and Pullman, B.

[22] Commoner, B. (1962) op. cit. also (1964), *Nature* 202, pp. 960–8. Also Dean, A. C. R. and Hinshelwood, Sir Cyril (1963), *Nature* 199, pp. 7–11, 201, pp. 232–9 and 202, pp. 1046–12.

[23] Quastler, H. (1959), *Nature*, p. 228.

[24] Dixon, M. and Webb, E. C. (1964), *Enzymes*, 2nd ed., see p. 496 'Ligases' are enzymes which bring about the synthetic linking together of two molecules, simultaneously with breakdown of ATP. The energy liberated by the breakdown serves for the synthesis. The importance, and even the very existence, of ligases was only recently recognized.

As to the second point raised by the quotation from Dixon and Webb (p. 522), it may be added that as long as the possibility remains open (as it clearly does at present) that life may be able to traverse space in certain forms, the discovery of life on other planets need not affect the issue.

[25] Elsasser (1958), op. cit., p. 159. Also MacKay, D. M. (1958), *Aristot. Soc. Suppl.*, 32, pp. 105–22.

[26] Whyte, L. L. (1957), *Brit. J. Philos. Sci.*, 7, pp. 347–50, argues that the nature of quantum mechanics is such that the principles of organic ordering cannot possibly be deduced from them alone. He supposes, however, that the discovery of more general structural principles and more powerful approximated methods may eventually enable physical theory to provide a satisfactory representation of the principles characterizing organic processes.

[27] Polanyi, M., (1962), op. cit.

[28] Elsasser, W. M. (1962), *J. Theoret, Biol.*, 3, p. 164.

[29] Polanyi, M. (1962), op. cit.

[30] Burnet, Macfarlane (1956), *Enzyme, Antigen and Virus*, p. 163.

[31] Commoner, B. (1962), op. cit., pp. 331 and 334.

[32] The success of the DNA hypothesis in accounting for coding in the germ plasm has led many people to assume that a similar mechanism might be used to explain the coding of memories in the central nervous system, including the brain of man. Here almost all is as yet speculation, with hardly any firm facts to go upon, and stupendous difficulties to be overcome. Perhaps one is particularly worth mentioning: It is assumed, indeed it seems to be inevitable, that the nucleotid sequence in the RNA responsible for the memory must be in the first place established as a consequence of ionic flux induced by nerve impulses coming from the sensory centres. That is, in some way the sense organs and

their connectives must at some point in the process be instructing the cells where the memory is stored as to what particular nucleotide sequences they must retain and propagate if they are to do their job. The difficulty here is to explain how an electrical current could induce a molecular rearrangement which is thereafter irreversible and immune to further perturbations of its electrical surround. The only possibility seems to imply the view that all possible RNA nucleotid sequences – all 10^{20} of them, and their correspondingly coded proteins – are already available within the cell; and that an impinging pattern of electrical impulses might select or re-orient some of these molecules at the expense of others. This seems to involve a requirement so staggering as, on our present knowledge, to rule out serious consideration of the mechanism altogether. Moreover, in my view it rules out, at least for the time being, any serious consideration of the view that nucleotid sequences are the basic mechanism for information storage in the nervous system. To attempt to explain further difficulties here would take too much space and involve too much technicality. Those interested are referred to my paper 'Ethology and the coding problem in germ cell and brain', *Zeit. für Tierpsychologie*, 20, pp. 529–51, 1963.

[33] Popper, K. R. (1950), *Brit. J. Philos. Sci.*, Vol. **I**.

[34] MacKay, D. M. (1961), *The Science of Communication – a Bridge between Disciplines*. Inaugural lecture in Univ. College of North Staffordshire, Keele.

[35] MacKay, D. M. (1966), 'Cerebral Organization and the Conscious Control of Action', in *Brain and Conscious Experience*. ed. Eccles, J. C. New York and Berlin.

[36] Broad, C. D. (1937), *The Mind and its Place in Nature*, pp. 76–7, London.

[37] Russell, Bertrand (1935), *Religion and Science*, London.

[38] This applies to the animal kingdom. In the case of plants it is much more questionable. Dr G. C. Evans has pointed out to me a number of important considerations in connexion with the evolution of plants which are relevant here. One of the great problems is posed by the fact that the biochemical characteristics of higher plants are *in addition to* the morphological and floristic characters, e.g. alkaloids in the Solanaceae. These are certainly not biochemical accidents in the sense that the occasional occurrence of haemoglobin in insects is a biochemical accident. Alkaloids are highly complex biochemical substances and are extremely difficult of synthesis. Far from being a simple biochemical accident their occurrence must be the result of a very elaborate series of highly controlled processes. The argument that many of these substances may have, or have had, a protective effect is based on very little evidence. In many cases it seems impossible to have any confidence in it at all. Thus the plants are on the

whole characterized by being able to produce whole series of peculiar compounds. It is probably this biochemical originality which has been the driving force in the evolution of the plants, a driving force which the insects specialized to particular plants have been forced to follow. Heterospory in the Pteridophyta is a supreme example of the 'originality' of plants. There is no conceivable function of heterospory in the Pteridophyta yet one of its types has led to the evolution of seed plants. In connection with heterospory, structures of fantastic similarity have developed independently, e.g. the carboniferous Lycopodiales and the Filicales.

In plants it is the highly elaborate ones which have the small chromosome numbers. In ferns chromosome numbers tend to be in the hundreds, in the Labiatae and the Compositae 5, 6 or 7. The chromosome number of Drosophila is 4 and there is no genetic evidence whatever for genes being present which differentiate say Drosophila from Calliphora. One is forced to the conclusion that the characteristics of the Diptera and of the family groups such as Muscidae and the Drosophilidae must be coded in an extra-nuclear mechanism.

Another extraordinary problem is posed by the main algal groups; for there is no adaptive significance in the presence of the brown pigment which characterizes the brown algae and distinguishes them from the green algae. Yet a whole host of morphological and other characters is related to these colour differences and the original early classification based on colour has stood the test of time and shown itself to be a natural classification. Here there would seem to be no selective mechanism whatever ensuring that the colour characteristics of these main groups of plants are maintained. Whatever mechanism it is must be completely and absolutely resistant to all destructive circumstances, e.g. thermal noise, independently of the protecting effect of natural selection.

[39] Raven, C. E. (1962), *Teilhard de Chardin: Scientist and Seer*, London.

[40] Whitehead, A. N. (1933), *Adventures of Ideas*, Cambridge.

[41] Teilhard de Chardin (1916), *La Vie Cosmique*, Paris.

[42] Quoted by C. E. Raven, (1962), op. cit., p. 184.

[43] Ryle, G. (1949), *The Concept of Mind*, London.

[44] Waddington, C. H. (1961), op. cit.

[45] Thorpe, W. H. (1961), *Biology, Psychology and Belief*, Cambridge.

[46] Apter, M. J. and Wolpert, L. (1965), *J. Theor. Biol.*, 8, have attacked the use of 'information' adumbrated by Quastler and adopted in these pages. But, as far as I can understand them, they miss or misrepresent some important points, so that their criticism fails in the main. It was unfortunate that they had not seen Quastlers 1964 book (p. 157 above, reference 33) before their paper was published.

[1] Huxley, Julian (1962), 'Higher and Lower Organization in Evolution'. *J. Royal College of Surgeons of Edinburgh*, **7**, pp. 163–79.

[2] Thouless, R. H., (1964), *Mind and Consciousness in Experimental Psychology*. Cambridge.

[3] Owen, A. R. G., (1964), *Can we Explain the Poltergeist?* New York.

[4] Pratt, J. G. (1964), *Parapsychology*, New York.

[5] Pratt, J. G. *et. al.* (1962–4), *J. of Parapsychology* (various papers) Ryzl, M. and Ryzlova, J. (1962), A case of high scoring in ESP performance in the hypnotic state, *J. Parapsychology*, **26**, pp. 153–71. Ryzl, M. and Pratt, J. G. (1963), A further confirmation of stabilized ESP performance in a selected subject. *J. Parapsychology*, **27**, pp. 73–83.

[6] Cutler (1960), *Psychological Reports*, **6**, pp. 283–9, has presented the hypothesis that one aspect of the evolution of intelligence is a consequence of the retention in higher species of a neotinic behavioural plasticity characteristic of primitive organisms during immature stages. This would suggest that sometimes an advance can come more readily by developing the less specialized characteristics of the immature form rather than by building upon the more specialized but sometimes more rigid characteristics of the adult.

[7] Clark, W. E. LeG. (1958), 'The Study of Man's Descent', in *A Century of Darwin*, ed. Barnett, S. A., London.

[8] Chance, M. R. A. (1962), in *Social Life of Early Man*, ed. Washburn, S. L., p. 29.

[9] Vallois, H. V. (1962), 'The Social Life of Early Man; the evidence of skeletons, in *The Social Life of Early Man*, ed. Washburn, S. L., London.

[10] Washburn, S. L. and DeVore, I (1962), 'Social Behaviour of Baboons and Early Man', in *Social Life of Early Man,* ed. Washburn, S. L., London.

[11] Goodall, J. M. (1964), *Nature*, **201**, pp. 1264–66.

[12] There are also convincing records of tool-using amongst birds and in at least one sub-primate mammal. As for birds, the Satin Bower Bird uses fibrous material as a brush with which to paint the sticks of its bower with a colouring material from berries or with charcoal, and one cannot possibly deny that this is

tool-using in the full sense. There is another bird which as a tool-user is so able as almost to stand in a category by itself; this is the Galapagos Woodpecker Finch which feeds on insects which inhabit the crevices in and under the bark of various cacti. Its bill is not long enough to reach the insects unaided and it does not normally attempt to do so. Instead it picks up a long cactus spine and pokes about with it at random in the crannies and crevices which the insects inhabit. Sooner or later an insect or spider runs out, whereupon the bird drops the spine and seizes the insect. Then another spine is picked up and the procedure is resumed. Both these cases clearly necessitate the utilization of the mechanical properties of an object as an extension of the bill of the bird. The California sea otter is in its way every bit as remarkable as a tool user as is the chimpanzee. The animal dives to the sea bottom and returns to the surface carrying a small boulder or a large stone and one of the large 'shellfish' such as abalone. The otter then floats on its back in the water, resting the stone on its chest, and hammers the mollusc on to the stone until the shell is broken and the meat can be devoured. It then drops the stone and returns to the bottom for another mollusc and comes up again with the same or another stone. The whole process has been carefully analysed and it is quite clear that there is intentional selection of a stone for the purposes of a tool and that all the behaviour shows striking evidence of at least elementary planning and foresight. An account of these and other cases will be found in Thorpe, W. H. (1962), *Learning and Instinct in Animals, 2nd ed.,* London. For the latest study of the sea otter, see Hall, K. R. L. and Schaller, G. B. (1964), *J. of Mammalogy,* 45. p. 287.

[13] Chance, M. R. A. in Washburn (1962), op, cit.

[14] The amygdala is one of the primary olfactory centres of the primitive hemispheres. The expanding cortex overflows it and there is evidence that the amygdala mediates such semi-visceral activities as chewing, drinking and salivation. Olfactory impressions of course have a powerful influence on visceral activity but there is recent evidence (Chance and Mead (1953), Symp. No. 7 *Soc. exp. Biol.*) that the enlargement of those parts of the cortex closely related to the amygdala is in some way connected with competition for dominance. It seems to have some ability to control emotive expression at high levels of social excitement and in so far as this faculty is dependent upon an enlarged amygdala, would also be a predisposing mechanism of 'tameness' which is assumed is a basic control faculty for the emergence of social life in hunting communities. Washburn, S. L., op. cit. (1962).

[15] Goodall, J. M. (pers. comm.)

[16] See also Dobzhansky, T. (1962), op. cit.

[17] Thorpe, W. H. (1961), *Bird Song: the biology of vocal communication and expression in birds,* Cambridge.

[18] Discussion of the evidence for these and other examples of cultural tradition in animals will be found in Thorpe, W. H. (1962), op. cit. (p. 164 above).

[19] Myiadi, D. (1959), On some new habits and their propagation in Japanese monkey groups, *Proc.* 15th Int. Congr. Zool., London, pp. 875–60.

[20] Dobzhansky, T. (1962), op. cit.

[21] Schultz, A. H. (1962), in Washburn, S. L., op. cit.

[22] Bergounioux, F. M. (1962), *Notes on the Mentality of Primitive Man,* in Washburn, S. L., op. cit., of which the next few lines are a paraphrase.

[23] Doubtless some of the proponents of the theory that learning in flatworms can be acquired through cannibalism will sooner or later hail this as providing support for their opinions!

[24] An extraordinary instance of this kind of behaviour in a chimpanzee is provided by Tinkelpaugh. Details will be found in Zuckerman. S., *Functional affinities in man, monkeys and apes* (1933), London, p. 154. The third baby of a chimpanzee (Mona), one of the colony of apes in the Department of Psychobiology at Yale University, died within twenty-four hours of its birth. The mother guarded the corpse jealously, and for a month resisted all efforts to remove it, carrying the decomposing remains wherever she moved. A short time before it was taken from her, she was seen to crack the skull open with her teeth and eat some of the contents. The significance of this extraordinary behaviour is entirely to seek; if it is found to be common it should perhaps be regarded as the forerunner of the remarkable human cults here being discussed.

[25] Crisler, L. (1959), *Artic Wild,* London.

[26] Tinkelpaugh, O. L. (1928), The self-mutilation of a male *Macaca rhesus* monkey, *J. Mammalogy,* **9**, pp. 293–300.

[27] Zuckerman, S. (1932), *Social Life of Monkeys and Apes,* London.

[28] For instance, papers by Bowlby, Kaufman and Kubie, in Bowlby, J. (1952), *Prospects in Psychiatric Research,* ed. J. M. Tanner, Oxford.

[29] See Fletcher, R. (1957), *Instinct in Man,* London, pp. 213 and 204.

[30] See Fletcher, R., op. cit., p. 199.

[31] Freud, S. (1949), *An Outline of Psychoanalysis,* London.

[32] Fletcher, R. (1957), op. cit., pp. 240 and 253.

[33] Dobzhansky, T. (1962), op. cit., p. 61.

[34] Gesell, A. and Amatruda, B. S. (1941), *Developmental Diagnosis*, New York.

[35] Dobzhansky, T. (1962), op. cit.

[36] Williams, H. A. (1962), 'Theology and Self-awareness', in *Soundings*, ed. Vidler, A. R., Cambridge.

[37] Lorenz, K. Z. (1963), *Das sogennante Böse: Zur Naturgeschichte der Aggression*, Wien.

[38] Maynard Smith, J. (1964), 'Group Selection and Kin Selection', *Nature*, **201**, pp. 1145–7, has, as a geneticist, given reasons for thinking that group selection can only be an effective method in the evolution of adaptation if the members of the group on which the selection is operating have a degree of kinship above a certain level.

[39] Bally, G. (1945), *Vom Ursprung und von der Grenzen der Freiheit*, Basel.

[40] Rensch, B. (1957), Äesthetische Faktoren bei farb- und formbevorzugungen von Affen. *Z. Tierpsych*, **14**, pp. 71–99.

[41] Thorpe, W. H. (1961), *Bird Song: the biology of vocal communication and expression in birds*, Cambridge.

[42] Hall-Craggs, J. (1962), The development of song in the blackbird *Turdus merula Ibis*, **104**, pp. 277–300.

[43] Bridges, Robert, *The Testament of Beauty*, Book **3**, 1, 393.

[44] Szöke, P. (1962), Zur Entstehung und Entwicklungsgeschichte der Musik. *Studia Musicologica* **2**, pp. 33–85.

[1] Marler, P. (1959), Developments in the study of animal communication'. Ch. 4 in *Darwin's Biological Work: some aspects reconsidered*, Ed. P. R. Bell, Cambridge.

[2] See p. 91 above.

[3] References to the chief papers by Otto Koehler and summaries of his work will be found in Thorpe (1962), op. cit.

[4] Lögler, P. (1959), Versuche zur frage de 'Zahl'-vermögens an einem Grau Papagai. *Z. Tierpsychol*, **16**, pp. 179–217.

[5] Thorpe, W. H. (1962), *op. cit.*, pp. 392–3.

[6] Diamond, A. S. (1959), *The History and Origin of Language*, London.

[7] Hinshelwood, C. (1960), Presidential Address *Proc. Roy. Soc.* (B), **153**, pp. 145–54.

[8] Rowell, T. E. (1960), On the retrieving of young and other behaviour in lactating golden hamsters. *Proc. Zool. Soc.*, **135**, pp. 265–82.

[9] Even dogs will occasionally attack apparently loved masters or mistresses; though it nearly always transpires that there is some hindrance to normal recognition. See Grzimek, B. *Z. Tierpsychol* (1953 and 1954), **10**, pp. 71–6 and **11**, pp. 144–9.

[10] Kirkman, F. B. (1937), *Bird Behaviour: a contribution based chiefly on a study of the Black-headed Gull*, London.

[11] Lack, D. (1939), 'The Behaviour of the robin'. *Proc. Zool. Soc. Lond.*, **109**, pp. 169–78.
At the time of going to press particulars of some very remarkable work on the concept forming abilities of the pigeon has been published by Herrnstein and Loveland, *Science* (1964), **146**, pp. 549–51. Pigeons were trained to respond to the presence or absence of human being in photographs. The precision of their performances and the ease with which the training was accomplished suggest greater powers of conceptualization than are ordinarily attributed to animals. Unless there is something extraordinary about the conceptual capacities of pigeons, these findings show that an animal readily forms a broad and complex concept when placed in a situation that demands one. The results are so remarkable in view of the difficulty that animals, and even primitive human

beings have in understanding complex pictures, that one will wait with keen interest for confirmation. It must, however, be said that the methods employed were highly sophisticated and every care taken against misinterpretation. The graphs published summarize a large mass of data. See also Rensch, B. (1962), *Gedächtnis, Abstraktion und Generalisation bei Tieren*. Köln.

[12] Lorenz, K. (1963), *Das sogenannte Böse :zur naturgeschichte der Aggression,* Vienna, p. 416. See Ch. V, especially pp. 108–12.

[13] Cary, J. (1958), *Art and Reality*, London.

[14] Blanc. A. C. (1962), 'Some evidence for the Ideology of Early Man', in *Social Life of Early Man*, ed. Washburn, S. L., London.

[15] Washburn, S. L. (1962), op. cit., pp. 119–20.

[16] See *Africana* (1964), **2** (No. 1), p. 33 (Nairobi).

[17] Habgood, J. S. (1962), 'The uneasy Truce between Science and Theology', in *Soundings*, ed. Vidler, A. R.; Cambridge.

[18] Marett, R. R. (1908), *The Threshold of Religion*, London.

[19] Belief in metempsychosis is known in a number of primitive cultures today. Among the Eskimo, grandparents used to name the young child and if it was the first-born, raise it as their own. The names were often those of deceased relatives who had been good men and women in their day and whose spirit was thought to live in the body of the child so named. Beliefs such as this can have some curious side-effects. In the case of the Eskimo, one of the results was that it was thought wrong to chastise children because you were really spanking a deceased grandparent – which was thought to be inadvisable for various reasons. With the waning of the belief in metempsychosis, partly due to the teaching of the missionaries, some Eskimo families now take another line with naughty children and spank them freely – having come to the conclusion that spanking is definitely a good thing (see Wilkinson, D. (1956), *Land of the Long Day*, London, p. 236).

[20] 'Mana' implies primarily the belief of primitive peoples that certain physical objects are the vehicles of occult powers. Some cultures believe this kind of power is held also to belong to the king or governing class. The sanction behind tabu is usually the mana of the chief or his lineage.

[21] Thoday, J. M. (1958), 'Natural Selection and Biological Progress', in *A Century of Darwin*, ed. Barnett, S. A., London.

[22] Dobzhansky, T. (1962), op. cit., p. 325.

[32] Ginsberg, M. (1944), *Moral Progress*: being the Fraser Lecture in the University of Glasgow, for 1944, Glasgow.

[24] Waddington, C. H. (1961), op. cit., p. 104.

[25] Again in education there are two outstanding requirements: to educate for change not for stability; and to educate for leisure not long hours of work. For change and more leisure are inevitable corollaries of the population explosion coupled with technological advance and automation, as Dr Mongar recently pointed out very graphically (*pers. comm.*).

[26] Bridges, Robert (1929), *The Testament of Beauty*, Oxford (Book 2, **1**, 204–10).

[27] Muller, H. J. (1950), 'Our Load of Mutations'. *Amer. J. Human Genetics*, **2**, pp. 111–76.

[28] Fisher, R. A. (1930), *The Genetical Theory of Natural Selection*, Oxford.

[29] Richter, C. P. (1959), 'Rats, Man and the Welfare State'. *Amer. Psychologist*, **14**, pp. 18–28.

[30] Dobzhansky, T. (1962), op. cit.

[31] Muller, H. J. (1935), *Out of the Night: a biologist's view of the future*, New York. Muller, H. J. (1959), *The Guidance of Human Evolution: Perspectives in Biological Medicine*, **3**, pp. 1–43.

[32] Ginsberg, M. (1944), op. cit., p. 17.

[33] I have deliberately said little about the population problem, frightening though it is. The need for action is at last coming to be widely recognized and, although there is an immense amount of research still required, methods are coming fast. Lord Brain and Professor A. S. Parkes have discussed the subject recently, 1964–5, *Advancement of Science*, **21**, pp. 221–9 and 509–13. They urge first the spread of precise demographic information by an international body so that every nation would have, in the clearest possible terms, the information on which to base its own population policy. Secondly, there is the acute social problem of motivation. Man through selection has been endowed with an urge to mate and with a parental urge at a level appropriate to a low survival rate. Those concerned with the many more personal ethical problems involved, either centrally or marginally, with the population explosion are recommended to read *The Sanctity of Life and the Criminal Law* by Glanville Williams, 1958, London.

CHAPTER 5

[1] Thoday, J. M. (1958), op. cit.

[2] Habgood, J. S. (1962), op. cit.

[3] Root, H. E. (1962), in Vidler, op. cit., quoted on p. 12.

[4] Haldane, J. B. S. (1927), *Possible Worlds*, London, p. 240.

[5] Cary, J. (1958), op. cit.

[6] Hinshelwood, C. N. (1960), Presidential address to the Royal Society *Proc. Roy. Soc.* (B), **151**, pp. 300–7.

[7] Sciama, D. W. (1959), *The Unity of the Universe*, London.

[8] Bondi, H. and Gold, P. J. (ref. in Sciama, op. cit.).

[9] Langer, S. (1951), *Philosophy in a New Key*, New York.

[10] Thorpe, W. H. (1961), *Biology, Psychology and Belief*, The 14th Eddington Memorial Lecture, Cambridge.

[11] Hobhouse, L. T. (1906), *Morals in Evolution*, See specially the 7th edition, 1950, with an introduction by Professor Morris Ginsberg.

[12] Huxley, J. S. (1947), *Evolution and Ethics*, London.

[13] Waddington, C. H. (1961), op. cit.

[14] See also Maynard Smith, J. (1964), 'Group Selection and Kin Selection', *Nature* **201**, pp. 1145–7.

[15] Robinson, J. A. T. (1963), *Honest to God*, London.

[16] Woods, G. F. (1962), 'The Idea of the Transcendent', in *Soundings*, ed. Vidler, A. R., Cambridge.

[17] To state the problem of evil in this way does not of course to any degree provide an answer to it. It would be neither within the scope of this book, nor within my competence to attempt such. I shall therefore make no further reference to it except to point out that to my mind the best statement of the Christian answer to the problem of evil has been provided by one whom many theological critics castigated for ignoring the problem: I refer to Teilhard de Chardin. Those who wish to learn his views on this should read *Le Milieu*

Divin, especially the third chapter of Part II and the following section which concludes both Parts I and II.

[18] Dante Aligheiri. *De Monarchia* I, viii (quoted by Charles Williams, *The Figure of Beatrice* (1958), London.)

[19] Dante Aligheiri *Il Paradiso*, XX, lines 130–32, and XXXIII, lines 85–90.

[20] Williams, H. A. (1962), *Theology and Self-awareness,* in Vidler, op. cit,

[21] Woods, G. F. (1962), op. cit.

[22] Malcolm, N. (1958), *Ludwig Wittgenstein : a Memoir*, Oxford.

[23] See Baillie, J. (1962), *The Sense of the Presence of God*, London.

[24] Smart, R. N. (1962), *The Relation between Christianity and the other Great Religions*, in Vidler, op. cit., Cambridge.

[25] Newman, J. H. (1870), *Grammar of Assent*.

[26] Raven, C. E. (1962), *Teilhard de Chardin : scientist and seer*, London.

[27] Besant, J. (1963), in *Objections to Christian Belief*, ed. Vidler A. R. Cambridge.

[28] Besant, J. (1963), op. cit.

[29] Baillie, J. (1962), op. cit.

[30] Webb, C. C. J. (1919), *God and Personality*, London.

[31] Montefiore, H. W. (1962), 'Towards a Christology for Today', in *Soundings,* ed. Vidler, A. R., Cambridge.

[32] Besant (1963), op. cit., p. 104.

[33] Woods, J. S. (1962), op. cit.

[34] A splendid disquisition on *The Vision of Nature* was given us in 1961 by Sir Cyril Hinshelwood in his Eddington Lecture (Cambridge University Press).

[35] Dante Aligheiri, *Il Paradiso*, Canto XXXIII, lines 52–81.

[36] Arber, A. (1957), *The Manifold and the One*, London.

[37] Raven, C. E. (1927), *The Creator Spirit*, p. 208, London.

[38] Inge, W. R. (1925), *In Science, Religion and Reality*, ed. Needham, J., London, see p. 385.

[39] Underhill, E. (1911), *Mysticism*, London, see p. 431.

[40] Zaehner, R. C. (1936), *Mysticism sacred and profane,* London.

[41] Broad, C. D. (1924), *Contemporary British Philosophy*, Vol. 1, ed. Muirhead, London.

[42] See also Cary, J. (1958), op. cit.

[43] Polanyi, M. (1958), *Personal Knowledge,* London.

[44] Dobzhansky, T. (1962), op. cit., Muller quoted on p. 50, see also pp. 320-21.

[45] Hardin, G. (1959), *Nature and Man's Fate*, New York.

[46] Russell, B. (1935), op. cit., p. 81.

[47] Haldane, J. B. S. (1932), *The Inequality of Man.*

[48] See Vidler, A. R. (1962), op. cit., p. 145 for the following statement: 'By refusing to claim absolute authority for itself, the church witnesses to the continuing activity of the Holy Spirit who is ever guiding it into a fuller understanding of that truth which will be known completely or absolutely only "at the last day".'

[49] Polanyi, M. (1958), op. cit.

[50] Raven, C. E. (1962), op. cit.

[51] Teilhard de Chardin, P. (1964) *The Future of Man*, London. (Translated by Norman Denny.)

[52] This may appear superficially similar to the evolutionary introgression known in a number of invertebrate groups. In fact it is, of course, basically different from it in regard to the selective mechanisms at work.

Index

175